Chemical Safety Handbook

FOR THE SEMICONDUCTOR/
ELECTRONICS INDUSTRY

David G. Baldwin, CIH
Hewlett-Packard
Palo Alto, California

Michael E. Williams, CIH
Apple Computer, Inc.
Cupertino, California

Patrick L. Murphy, CIH
Intel Corporation
Chandler, Arizona

Chemical Safety Handbook

SECOND EDITION

FOR THE
SEMICONDUCTOR/
ELECTRONICS
INDUSTRY

OEM Press   Beverly, Massachusetts

Copyright © 1996 by David G. Baldwin

Second Edition

ISBN 1-883595-07-X

Printed in the United States of America

All rights reserved. No part of this book may be reproduced in any form or by any electronic or mechanical means, including information storage and retrieval systems, without written permission in writing from the publisher, except by a reviewer who may quote brief passages in a review.

Questions or comments regarding this book should be directed to:

OEM Health Information, Inc.
Suite 814
181 Elliott Street
Beverly, MA
01915-3080

(508) 921-7300
(508) 921-0304 (Fax)

OEM Press® is a registered trademark of OEM Health Information, Inc.

Contents

Preface		vii
Acknowledgments		ix
1.0	Introduction	1
2.0	Definitions	2
3.0	Emergency Procedures	10
	3.1 Emergency Notification	10
	3.2 Emergency Response	10
	3.3 First Aid	11
4.0	Hazard Control	13
5.0	Personal Protective Equipment (PPE)	14
	5.1 General Requirements	14
	5.2 Acid and Solvent Gloves	15
	5.3 Respirators	16
6.0	Maintenance Considerations	17
	6.1 Confined Space Entry	18
	6.2 Opening Equipment	18
	6.3 Cleaning Arsenic Residues	19
	6.4 Soldering Systems	19
	6.5 Plasma Etchers	20
7.0	Chemical Handling Procedures	22
	7.1 General Requirements	22
	7.2 Corrosives (Acids and Bases)	23
	7.3 Solvents (Including Resists)	24
	7.4 Oxidizers	26
	7.5 Process Gases	27
	7.6 New Chemicals	28

8.0	Radiation Safety	29
	8.1 General Radiation Information	29
	8.2 Radioactive Materials	30
	8.3 Radiation Machines	30
	8.4 Ultraviolet (UV) Light	32
	8.5 Radiofrequency/Microwave (RF/MW) Radiation	33
	8.6 Lasers	34
9.0	Safety and Health Information	34
	9.1 Medical and Exposure Records	34
	9.2 Material Safety Data Sheets	35
	9.3 Additional Information	46
10.0	Trade-Name Compounds and Chemical Mixtures	47
	10.1 Photoresists	47
	10.2 Generic Components	48
11.0	Chemical Hazards and Precautions	107
	11.1 Format	107
	11.2 Generic Chemicals	111
12.0	References	420
13.0	Index for Chemical Hazards and Precautions	423

Preface

This handbook is primarily intended for the use of operators and technicians working in the semiconductor/electronics industry. It provides information on safety procedures and chemicals used in the semiconductor/electronics industry. It should be used as part of a comprehensive hazard communication program which includes employee training, material safety data sheets, and container labeling. This book is not a substitute for chemical training and does not include information on general safety.

Certain procedures in this handbook apply only to semiconductor/electronics manufacturing. Also, some of the hazard information applies only to chemicals used in concentrations and processes specific to typical semiconductor/electronics operations. Therefore, particular care is necessary when applying information in this handbook to other industries or to new or unique operations.

A balance was sought between keeping the book at a reasonable size and providing enough detail to adequately identify major hazards and precautions associated with chemicals used in the semiconductor/electronics industry. Individuals using this book should receive training on specific hazards particular to their operations. Also, variations in a facility's operations or process may require deviations from procedures outlined in this book. Contact your supervisor or safety department for information on these appropriate variations.

Acknowledgments

We are indebted to George Rakonitz, Vice-President of Commercial Relations, National Semiconductor Corporation, for allowing the use of National Semiconductor's *Chemical & Radiation Safety Handbook*, fourth edition, as a basis for the first edition of this book. National Semiconductor's handbook was the culmination of considerable effort by many people over many years.

The first edition of the *Chemical Safety Handbook* was also the product of two other co-authors, Andy McIntyre, Environmental and Occupational Risk Management, and Lewis Scarpace, Intel. Unfortunately, the demands of their day jobs prevented them from working on the second edition.

For their suggested improvements to the second edition, we would like to thank Ron Cox, Harris Semiconductor, and Mike Montalvo, Sony Microelectronics. Their recommendations, along with those we received from our colleagues at work, helped enhance this edition.

— *The Authors*

The information and recommendations contained in this publication have been compiled from sources believed to be reliable and to represent the best opinion on the subject as of January 1996. Because this book is intended to be a useful summary of information and not an exhaustive compendium, it may not include information which is important for a particular user. The primary reference sources used are listed in the bibliography. The reader should consult these sources in order to obtain a more complete explanation of the hazards, precautions, and other information contained in this book. The publisher makes no warranty, guarantee, or representation as to the correctness or sufficiency of any information or recommendations herein. This book is intended to be used as part of a comprehensive hazard communication program which includes employee training, reference to actual material safety data sheets, container labeling and ongoing monitoring of research, and new data on chemical and radiation hazards. Read the preface for additional information before using this book.

1.0 Introduction

This handbook is intended as a readily available resource for information on the many chemicals used in the semiconductor/electronics industry and procedures for their safe use. No one company uses all of the chemicals listed in this book. However, most major chemicals used by the electronics industry are included. Some information on radiation hazards is also contained in the book. Contact your supervisor or safety department if you need additional information on a specific chemical or radiation source.

The book is divided into 13 sections:

- Section 1 is the introduction.

- Section 2 gives definitions for terms used in the book.

- Sections 3 through 9 cover general chemical safety information.

- Section 10 lists the components of some trade-named chemical mixtures used in the electronics industry.

- Section 11 lists hazards and precautions associated with generic chemicals used in the electronics industry.

- Section 12 lists references used in making this book.

- Section 13 is an index for Section 11; it shows the different names used for the generic chemicals and the page number where they can be located.

This reference guide is intended for the use of operators and technicians working in the semiconductor/electronics industry. It should be used as a complement to a comprehensive hazard communication program which also includes employee training, material safety data sheets, and container labeling. It is only a guide for the handling of chemicals used in semiconductor/electronics manufacturing. It is not a substitute for chemical training and does not include information on general safety.

2.0 Definitions

Acid Corrosive materials whose water solutions contain hydrogen ions (H^+). Like bases, in sufficient amounts these materials burn, irritate, or destructively attack organic tissues such as the skin, lungs, and stomach.

Acute Something lasting for only a short time. Acute exposures to chemicals occur over a short period of time (minutes or hours). Acute toxic effects usually occur rapidly as a result of short-term exposures and are often of short duration (e.g., headache, nausea, vomiting).

American National Standards Institute (ANSI) The American National Standards Institute is a voluntary organization that develops national consensus standards for a wide variety of devices and procedures.

Autoignition Temperature (AT) The temperature at which a gas or vapor can explode or burst into flames with no other source of ignition.

Base Corrosive materials whose water solutions contain hydroxyl ions (OH^-). Like acids, in sufficient amounts these materials burn, irritate, or destructively attack organic tissues such as the skin, lungs, and stomach.

Carcinogen A chemical that causes cancer. For the purposes of this book, a chemical is considered to be a carcinogen if it is regulated by the Occupational Safety and Health Administration (OSHA) as a carcinogen or is considered a human or animal carcinogen by the American Conference of Governmental Industrial Hygienists (ACGIH), the International Agency for Research on Cancer (IARC), or the National Toxicology Program (NTP).

Chemical Abstract Service (CAS) Number A unique number assigned to each chemical by the Chemical Abstract Service, a division of the American Chemical Society.

Chronic Something that lasts a long time, is permanent, or takes a long time to develop. Chronic exposures to chemicals occur over a long period of time (months or years). Chronic toxic effects generally occur as a result of long-term exposure and are of long duration (e.g., cancer, liver damage).

Combustible Liquids Materials with a flash point at or above 100°F (37.8°C).

Compressed Gas Any substance which, when enclosed in a container, gives a pressure reading of at least:

- 25 psig (pounds per square inch, gauge pressure) at 70°F, or
- over 89 psig at 130°F, or
- over 25 psig at 100°F for flammable materials.

Corrosive Any solid, liquid, or gas that irritates or destructively attacks organic tissues, such as the skin, lungs, and stomach. May also dissolve metal, concrete, and other materials.

Curie (Ci) A unit of radioactivity. Radioactive materials are measured in curies. One curie equals a thousand millicuries (mCi). One millicurie equals a thousand microcuries (µCi).

Explosive A chemical that causes a sudden, almost instantaneous release of pressure, gas, and heat when subjected to sudden shock, pressure, or high temperature.

Explosive Range (ER) The range of concentration of a flammable gas or vapor (% by volume in air) in which explosion can occur upon ignition in a confined area. Example: The ER of hydrogen is 4 to 75%. The smaller number is referred to as the lower explosive limit (LEL). The larger number is the upper explosive limit (UEL).

Flammable **Solid:** A solid other than a blasting agent or explosive which burns or ignites easily through friction, absorption of moisture, spontaneous chemical change, or retained heat from manufacturing or processing, or which can be ignited readily and when ignited burns so vigorously and persistently as to create a serious hazard.

Liquid: A liquid with a flash point below 100°F (37.8°C).

Gas: A gas that forms a flammable mixture with air at a concentration of 13% by volume or less, or a gas that forms a flammable mixture with air with a range wider than 12% by volume regardless of the lower limit.

Flash Point (FP) The minimum temperature at which a liquid gives off vapor in sufficient concentration to form an ignitable mixture with the air above the surface of the liquid.

Hazardous Production Material (HPM) HPM is a solid, liquid, or gas that has a degree-of-hazard rating in health, flammability, or reactivity of Class 3 or 4 as ranked by Uniform Fire Code Standard No. 79-3 and which is used directly in research, laboratory, or production processes which have as their end product materials which are not hazardous.

Health Hazard A chemical or radiation for which there is statistically significant evidence based on at least one study conducted in accordance with established scientific principles that acute or chronic health effects may occur in exposed employees.

Immediately Dangerous to Life or Health (IDLH) The IDLH concentration represents a maximum airborne concentration of a chemical from which, in the event of respirator failure, a person could escape within 30 minutes without any escape-impairing symptoms or any irreversible health effects. Unlike the TLV/PEL and STEL values, it is not a "safe" level of exposure.

Industrial Hygiene The science and art devoted to the recognition, evaluation, and control of those environmental stresses, arising in or from the workplace, which may cause sickness, impaired health and well-being, or significant discomfort and inefficiency among workers or among the citizens of the community.

Irritant A chemical which is not a corrosive but which causes a reversible inflammatory effect on living tissue by chemical action at the site of contact.

NFPA Diamond An identification system for hazards of materials established by the National Fire Protection Association. It addresses the health, flammability, reactivity, and related hazards that may be presented by short-term exposure to a material during handling under conditions of fire, spill, or similar emergencies.

Numbers from 0 through 4 are used in the top three subsections of the diamond to represent the degree of short-term health, flammability, and reactivity hazard. The bottom subsection is used for special hazard symbols.

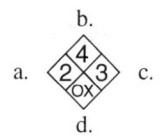

```
           b.
         ╱4╲
    a.  ╱2╳3╲  c.
         ╲OX╱
          ╲╱
           d.
```

a. Health Signal (color code–Blue): 4—materials that on very short exposure could cause death or major residual injury. 3—materials that on short exposure could cause serious temporary or residual injury. 2—materials that on intense or continued but not chronic exposure could cause temporary incapacitation or possible residual injury. 1—materials that on exposure would cause irritation but only minor residual injury. 0—materials that on exposure under fire conditions would offer no hazard beyond that of ordinary combustible material.

b. Flammability Signal (color code–Red): 4—materials that will rapidly or completely vaporize at atmospheric pressure and normal ambient temperature, or that are readily dispersed in air and that will burn readily. 3—liquids and solids that can be ignited under almost all ambient temperature conditions. 2—materials that must be moderately heated or exposed to relatively high ambient temperatures before ignition can occur. 1—materials that must be preheated before ignition can occur. 0—materials that will not burn.

c. Reactivity Signal (color code–Yellow): 4—materials that in themselves are readily capable of detonation or of explosive decomposition or reaction at normal temperatures and pressures. 3—materials that in themselves are capable of detonation or explosive decomposition or reaction but require a strong initiating source or which must be heated under confinement before initiation or which react explosively with water. 2—materials that readily undergo violent chemical change at elevated temperatures and pressures or which react violently with water or which may form explosive mixtures with water. 1—materials that in themselves are normally stable, but which can become unstable at elevated temperatures and pressures. 0—materials that in themselves are normally stable, even under fire exposure conditions, and which are not reactive with water.

d. Special hazard symbols are shown in the fourth space of the diagram or immediately above or below the entire symbol. Materials that demonstrate unusual water reactivity are identified with the letter W with a horizontal line through the center (W̶). Materials that possess oxidizing properties are identified by the letters OX.

Oxidizer A chemical other than a blasting agent or explosive that initiates or promotes combustion by spontaneously evolving oxygen either at room temperature or under slight heating. These chemicals can react with organic material or combustible liquids to cause or intensify fires.

Permissible Exposure Limit (PEL) The PEL is the maximum permitted 8-hour time-weighted average concentration of an airborne contaminant established by the U.S. OSHA. A few chemicals have a ceiling (instantaneous) concentration above which people should not be exposed. These PELs are identified with a "C."

In instances when the PEL conflicts with the American Conference of Industrial Hygienists' recommended Threshold Limit Value (TLV), the lower of the two values is listed in Section 11.2. PELs used for this book were those in effect in April 1995.

pH pH is a value used to represent the strength of an acid or base. pH values of 1 to 6 indicate acids, with 1 being a very strong acid and 6 being a very weak acid. Neutral materials, such as pure water, have a pH of 7. Bases have pH values of 8 to 13, with 8 being a very weak base and 13 being a very strong base.

Pyrophoric A chemical that will ignite spontaneously in air at a temperature less than or equal to 130°F (54.4°C).

Radiation Energy in the form of electromagnetic waves (e.g., light and x-rays) or energy emitted in the form of particles (e.g., alpha particles and beta particles). Radiation can be divided into two major categories: ionizing radiation and nonionizing radiation.

Radioactive Material Elements (either natural or artificial) that disintegrate over time. Energy in the form of ionizing radiation is emitted during the process. Radioactive material can be measured in curies.

Reactive Chemical Refers to unstable chemicals or chemicals that can form a hazardous situation when they come in contact with other chemicals (e.g., hazardous decomposition, hazardous by-products, hazardous polymerization).

Rem Rem (roentgen equivalent man) is a unit for measuring ionizing radiation. One rem equals 1000 millirems (mRem).

Sensitizer A chemical that causes a substantial proportion of exposed people or animals to develop an allergic reaction in normal tissue after repeated exposure to the chemical.

Short-Term Exposure Limit (STEL) A STEL is defined as a 15-minute time-weighted average (TWA) exposure which should not be exceeded at any time during a workday even if the 8-hour time-weighted average is within the TLV. The STEL is the concentration to which workers can be exposed continuously for a short period of time without suffering from adverse effects.

Solvent In common practice the term *solvent* is often used as a general synonym for most organic liquids.

A solvent is any substance capable of dissolving another substance (solute) to form a uniformly dispersed mixture (solution) at the molecular or ionic size level. Water is the most common solvent.

Threshold Limit Value (TLV) The TLV is an 8-hour time-weighted average (TWA) airborne concentration of a substance and represents conditions under which it is believed that nearly all workers may be repeatedly exposed day after day without adverse effect. TLVs are guidelines developed by the American Conference of Governmental Industrial Hygienists. A few chemicals have a ceiling (instantaneous) concentration above which people should not be exposed. These TLVs are identified with a ''C.''

In instances in which the TLV value conflicts with the U.S. Occupational Safety and Health Administration's Permissible Exposure Limit (PEL), the lower of the two values is listed in Section 11.2. TLVs used for this book were taken from the 1995–1996 TLV booklet.

Unstable (Reactive) A chemical which in the pure state, or as produced or transported, will vigorously polymerize, decompose, condense, or will become self-reactive under conditions of shocks, pressure, or temperature.

Vapor Pressure (VP) The pressure (in millimeters of mercury) characteristic of a vapor in equilibrium with its liquid or solid form. In this book the VP is given for 20°C unless otherwise noted.

The VP of a chemical relates to its speed of evaporation. Example: Since xylene has a higher VP than propylene glycol, it will evaporate faster. If equal quantities of xylene and propylene glycol were spilled, higher airborne concentrations of xylene would result.

Water-Reactive A chemical that reacts with water to release a gas that is either flammable or presents a health hazard.

3.0 Emergency Procedures

3.1 Emergency Notification

In case of an emergency, call the emergency number. You should NOT use a telephone that will require you to remain in a hazardous area. When the phone is answered you will be asked to give the following information:

- Type of emergency — gas leak, chemical spill
- Number and type of injuries (if any)
- Location of the emergency — building, floor, area
- Other areas affected (if any)
- Your name and the phone number you are calling from
- Location where you will meet the responders

Stay on the phone until you are told to hang up. Stay in or near the area of the emergency so you can provide additional information to the Emergency Response Team.

3.2 Emergency Response

Most semiconductor fabrication facilities and other large electronics manufacturing sites have an Emergency Response Team (ERT) specifically trained to handle emergencies. When you call the emergency number the ERT will respond to the problem. If the emergency requires assistance from outside professional emergency personnel such as the fire department or ambulance service, the emergency control station operator will request them (see Figure 1).

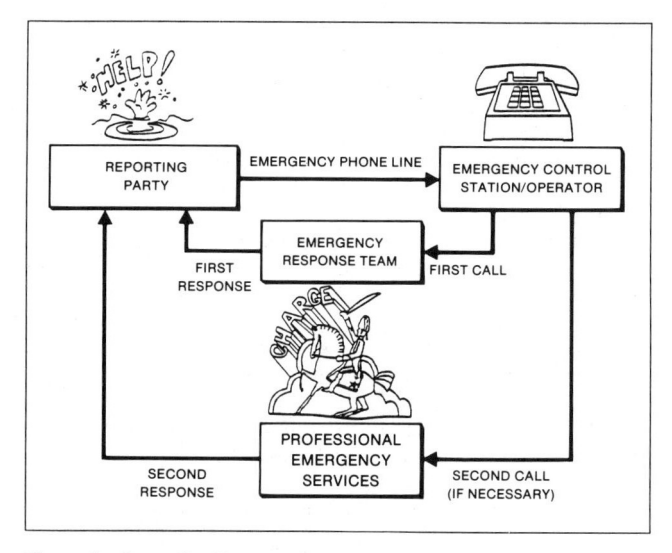

Figure 1. Reporting Emergencies

During an emergency, ERT members typically have the authority to order evacuations, remove unnecessary personnel from the emergency area, and assume general control. Sometimes the ERT shuts a system down for safety repairs. When this happens, the shut-down equipment should not be restarted without approval of the ERT leader or safety department.

3.3 First Aid

If skin or eyes are contacted by a corrosive liquid or organic solvent, the affected area must be immediately rinsed with large amounts of cold running water for 15 minutes or until medical assistance arrives. For eye contact and other major splashes, eye washes and safety showers should be strategically located throughout the work areas. Exposures to cryogenic gases or liquids should be rinsed with lukewarm water rather than cold water.

For chemical contact with the eyes, the affected person must hold both the upper and lower eyelids open with the thumb and forefinger during the rinse time. Flush the entire surface of the eye, especially the lower eyelid because the chemical will pool there.

If clothing is contaminated with chemicals, remove it while showering. After the contaminated clothing has been removed from the person, do not touch it without wearing the proper gloves. Do not reuse contaminated clothing until after it has been laundered. If there are any doubts as to the effectiveness of the laundering in removing the contaminant, contact the safety department for proper waste disposal.

Frequently there is a delay between skin contact with a chemical and the effects of exposure. For example, it can take up to 24 hours before the pain due to skin contact with hydrofluoric acid is first felt. Also, some chemicals are absorbed through the skin with little or no damage to the skin. Therefore, it is important to rinse immediately for 15 minutes and get medical attention whenever skin contact with any chemical is suspected. Do not rely on pain or other symptoms to indicate the need for first aid.

The area supervisor should be notified of all accidents and the employee should visit the plant nurse for appropriate treatment AFTER the 15-minute water rinse. If a plant nurse is not on duty, the supervisor should follow plant procedures AFTER the 15-minute water rinse to arrange for further medical care.

If exposure to dopants (e.g., arsine, phosphine, diborane) or acid gases is suspected, the affected employee(s) should immediately get medical attention for appropriate treatment.

4.0 Hazard Control

There are a variety of methods that can be used to control hazards in the workplace. These controls fall into four major categories. In order of general effectiveness, they are:

- Substitution
- Engineering Controls
- Administrative Controls
- Personal Protective Equipment

The preferred method for controlling hazards is to remove or reduce the hazard by changing the operation. If this is not practical the next best method is through engineering controls where the hazard and the person are separated. If this is not feasible, administrative controls should be used. Personal protective equipment should only be used when other control methods are not adequate to minimize the risk to the employee.

Example: The preferred method for controlling a hazardous exposure to methyl alcohol would be to remove or reduce the hazard by using water or isopropyl alcohol in place of the methyl alcohol. If this is not practical, the next best method would be to install local exhaust ventilation to remove the methyl alcohol vapors (an engineering control). If this is not feasible, the time a person spends cleaning with methyl alcohol could be limited. Respirators should only be used when other controls cannot eliminate the methyl alcohol overexposure.

In practice, a combination of controls is often employed such as using isopropyl alcohol, local exhaust ventilation, and solvent gloves and safety glasses.

5.0 Personal Protective Equipment (PPE)

5.1 General Requirements

Protective clothing should be worn by all employees who work with or around chemicals. Wash all reusable protective clothing with soap and water and check for holes and cracks in the material before each use. Throw away used solvent and acid gloves at the end of the shift. Store protective clothing in a location separate from chemicals.

Personal protective equipment (PPE) is provided by your company. For nonstandard items, contact the safety department.

- **Acid or Solvent Gloves** must be worn by employees who work with chemicals. Gloves should be leak checked before each use. Discard the glove if it fails the leak test or is wet on the inside. Used solvent and acid gloves must be thrown away at the end of the shift, even for operations where chemicals do not come in direct contact with the gloves. Gloves must not be shared between individuals.

- **Faceshields** should be worn in addition to safety glasses by employees using, pouring, or mixing chemicals.

- **Plastic or Rubber Aprons and Armguards** should be worn by employees using, pouring, or mixing chemicals. When not in use, aprons and armguards should be hung in an appropriate area to prevent damage and contamination.

- **Rubber Boots** that are chemically resistant should be worn by employees, such as plating operators, who work in wet environments. Steel- toed chemical boots may be required if the job involves material handling.

- **Safety Glasses** must be worn in all areas where chemicals are used. The safety glasses should be impact-resistant eyeglasses with sideshields. Safety glasses provide only limited protection against chemical splashes.

- **Safety Shoes** should be worn by employees who regularly lift or move heavy objects (e.g., moving gas cylinders, equipment, or solvent canisters).

- **Work Shoes** should be worn during all chemical operations. Work shoes are completely enclosed leather or simulated leather with low, flat heels. No cloth shoes or sandals should be worn when working with chemicals.

5.2 Acid and Solvent Gloves

Different glove types are required for corrosives and organic solvent handling. Often gloves made of nitrile butyl rubber (NBR) are worn when working with solvents, while latex rubber gloves are typically used when working with acids. Nitrile gloves should be worn when working with mixtures containing both acids and solvents (e.g., certain strippers). Some chemicals may require a specific type of glove different from what is generally used for its category (e.g., M-Pyrol). Check with your supervisor or safety department for information on what type of gloves to use for your operations.

In general, there is no such thing as "impermeable" plastic or rubber gloves. No one clothing material will be a barrier to all chemicals. For a given clothing material type, chemical resistance can vary significantly from product to product. For example, not all brands of nitrile gloves provide equal protection.

With most electronics manufacturing operations, solvent or acid gloves are used to protect against occasional splashes or minor surface contamination. However, for operations where the gloves are dipped into the liquid, the time it takes for the liquid to soak through the gloves becomes important. This time can be decreased if the chemicals are heated. Check with your supervisor or safety department when direct glove contact with chemicals is anticipated.

Acid and solvent gloves cannot be decontaminated and reused. Once a chemical becomes absorbed into the material, some of this chemical will continue to diffuse through the glove even after the surface has been decontaminated. Significant amounts of the chemical may reach the inside of the glove. This may not occur during the work shift but can take place while the glove is being stored overnight. The next morning when the person dons the glove, the hand may come into contact with the hazardous chemical. Because the chemical remains in the glove material even when not in contact with the liquid, solvent and acid gloves may not be reused.

5.3 Respirators

The type of respirator needed for a particular job varies with the chemical contaminant and its concentration in the air. For this reason it is important that all jobs where respirators are worn be evaluated by the safety department first.

Specific training on respirator use and fit testing of the respirator are necessary to ensure the respirator does not become a hazard from improper selection, use, or storage. A medical examination may also be necessary to ensure you are in the right physical shape to use a respirator. Contact your supervisor or safety department to schedule respirator training. Because this procedure takes time, call when you first anticipate the need for a respirator.

There are two major categories for respirators: air-purifying respirators (e.g., cartridge respirators) and air-supplying respirators (e.g., self-contained breathing apparatus — SCBA). Air-purifying respirators remove the contaminant before it is inhaled; they do not supply oxygen or protect against high concentrations of air contaminants.

Use of respirators during emergencies and for entry into confined spaces can be particularly hazardous. Specific training beyond what is required for typical respirator usage is needed for these activities.

Do not enter a confined space or use a respirator for emergency response without having successfully completed the specific training required for the job.

Respirators are not a substitute for proper engineering controls. Respirators may only be used to prevent harmful exposures:

- During the time necessary to install or implement other controls

- When other controls cannot adequately remove the contaminant at the source (e.g., during equipment maintenance)

- During operations in which there is significant potential for uncontrolled toxic gas leaks such as toxic gas cylinder changes

- In emergencies

Use a respirator only when you are authorized and trained to do so. You should use the respirator supplied by your company. Do not buy your own respirator.

6.0 Maintenance Considerations

Maintenance operations typically have the greatest potential for acute exposures to chemicals. Specific precautions and training are needed for each different maintenance operation. The information provided in this section is only an outline. It pertains to four maintenance operations that can be particularly hazardous if not performed properly. Contact your supervisor or safety department for specific information on these and other nonroutine tasks.

The four maintenance operations covered in this section are: (1) confined space entry, (2) opening contaminated equipment, (3) cleaning arsenic residues, and (4) cleaning soldering systems. You should closely follow the safety procedures your company has established for performing equipment maintenance and cleaning. There are many maintenance operations beyond the four mentioned here that have unique hazards associated with them.

6.1 Confined Space Entry

Before entering a confined space it is important that you receive training on the safety procedures required for this operation. Even with an air-supplied respirator, harmful exposures to toxic gases and liquids can occur from skin absorption of the chemical.

Explosion hazards can be a particular problem in confined spaces. The lack of air movement can allow vapors to accumulate until explosive concentrations are reached.

Even when hazardous chemicals are not present in confined spaces, oxygen deficiencies can be a health hazard. Oxygen deficiencies can occur for many reasons, such as the rusting of metal, the presence of organic material, the use of nitrogen purges, the lack of air movement, and the use of cutting torches. **Do not enter a confined space unless you comply with all safety procedures required by your company for confined space entry.**

6.2 Opening Equipment

Opening equipment which contains chemical residues or residual toxic gases can be particularly hazardous. These operations should be evaluated by your safety department to minimize the possibility of a harmful exposure. There is a concern over possible exposure to the original chemicals used in the equipment. Also, the original chemicals can react to form other compounds which deposit on the inside of the equipment or are suspended in the pump oil when other than dry pumps are used.

Because of their complex chemistries, opening of certain plasma etchers for maintenance is of particular concern. This issue is specifically addressed in section 6.5.

6.3 Cleaning Arsenic Residues

Residues containing arsenic require special attention. It is a regulated OSHA carcinogen and the permissible exposure limit (PEL) for arsenic is low: 10 micrograms of arsenic per cubic meter of air. Routine cleaning procedures can easily result in significant amounts of arsenic in the air. To minimize the amount of arsenic in the air, arsenic-contaminated surfaces should **never** be brushed, swept, or cleaned with compressed air.

Arsenic residues should be cleaned with a vacuum that has high-efficiency particulate (HEPA) filters or the contaminated surfaces should be gently wetted and cleaned with damp towels. Parts with heavy arsenic residues should be cleaned in an exhausted enclosure or glovebox.

Acids should **never** be used to clean arsenic residues. Acids, such as hydrochloric acid, can react with arsenic and form extremely hazardous concentrations of arsine gas.

6.4 Soldering Systems

Typical solder is made of 65% tin and 35% lead. Temperatures used for soldering are not high enough to cause significant amounts of airborne lead. The major health concern with soldering is the possibility of contaminating hands with lead from the solder and subsequent transfer to food, smoking materials, etc., resulting in ingestion of lead. This can be prevented by keeping the work area free of accumulations of solder and by washing your hands before eating, drinking, or smoking.

Irritation can occur from exposure to high concentrations of smoke that forms when flux contacts hot solder. The smoke may also cause an allergic sensitization in certain individuals. With small hand soldering operations the amount of smoke from the fluxes is small and typically not a cause of irritation. Local exhaust is needed to remove flux smoke from large soldering systems such as wavesolder machines. It may also be warranted when hand soldering is done extensively for long periods of time.

Significant exposure to lead can occur when soldering systems are cleaned if surfaces coated with solder are dry swept, cleaned with a wire brush, or heated with a blow torch. Solder residues or spilled dross should be cleaned with vacuums that have high-efficiency particulate filters, or the contaminated surfaces should be gently wetted and cleaned with damp towels. Parts with heavy solder accumulation should be cleaned in an exhausted enclosure or glovebox.

6.5 Plasma Etchers

A wide variety of chemicals are used in dry etching depending on the material being etched and the type of dry etcher used. The majority of these are chlorine- and/or fluorine-based chemicals.

Because dry etchers operate as closed systems, chemical exposure to the operators of the equipment typically does not occur while the system is closed. One rare exception to this is when the purge cycle for older batch etchers is not long enough to adequately remove the etchant gases. Brief but irritating exposures to fluorine compounds have been reported when the door to these etchers is opened. Normally this can be corrected by increasing the length of the purge cycle prior to opening the etch chamber door.

The major safety emphasis for dry etchers has focused on potential exposures to maintenance personnel working on the reaction chambers, pumps, and other associated equipment that may contain reaction product residues. With plasma etchers that have fluorine-based chemistries (e.g., polyetch), the reaction products are irritative. With plasma etchers that use chlorine-based chemistries (e.g., metal etchers), the concern is possible exposure to suspected cancer-causing chemicals as well as odors and irritation.

The complexity of plasma metal etchers and the difficulty in characterizing the odors associated with their maintenance has made them the subject of many investigations. Chemicals present in significant quantities that have been identified in studies of various dry etchers include hexachloroethane, cyanogen chloride, and hydrogen chloride. A large variety of other chlorinated and fluorinated compounds have also been identified in plasma etcher studies but typically these are present in smaller airborne concentrations.

Plasma etcher maintenance should always be done according to the manufacturer's recommended procedures and those established by your company. When chemical residues and odors are generated, particularly those from chlorine-based chemistries, these procedures often require that air-supplied respirators be used. Parts that give off odors are typically placed in a sealed container, such as a plastic bag, and transferred to an exhausted enclosure for cleaning. If odors are anticipated, operations where the contaminated surface cannot be removed, such as cleaning plasma etcher reaction chambers, are typically performed on the off-shift or other times when the area is evacuated of nonessential personnel.

Cold traps are an old technology that is still occasionally used with plasma etchers. Cold traps for plasma metal etchers can be particularly hazardous. Gases in the traps are condensed in liquid form and the specific chemicals in the mixture are often unknown. Particular care and training are needed when servicing cold traps.

7.0 Chemical Handling Procedures

7.1 General Requirements

Every chemical container, chemical work station, and chemical storage area should be labeled, tagged, or marked with the identity of the hazardous chemical(s) contained within and the appropriate hazard warning. Labels on incoming containers of hazardous containers should not be removed or defaced.

Warning labels are designed as a quick reference but they may not contain all necessary information. Additional information is contained in this book and in the Material Safety Data Sheets (MSDS) provided by the manufacturer (see Section 9.2).

Chemicals should be used as directed in process specifications and handled only by trained employees. No unattended open containers should be outside the confines of an air-exhausted hood or wet deck.

Table 1 is a guideline showing which types of chemicals may be stored together and which types of chemicals should be separated. Check with your supervisor or safety department for the specific storage requirements for your facility.

When pouring, mixing, or dispensing chemicals, it is a good safety practice to have a minimum of two persons in the area. Another person should be able to see or hear you if you are in an area with hazardous chemicals.

Table 1. Chemical Compatibility Matrix

x = not compatible ☐ = compatible (unless otherwise indicated)	Acids, Inorganic	Acids, Oxidizing	Acids, Organic	Alkalis (Bases)	Oxidizers	Poisons, Inorganic	Poisons, Organic	Water Reactives	Organic Solvents
Acids, Inorganic			X	X		X	X	X	X
Acids, Oxidizing			X	X	X	X	X	X	X
Acids, Organic	X	X		X	X	X	X	X	
Alkalis (Bases)	X	X	X				X	X	X
Oxidizers		X	X				X	X	X
Poisons, Inorganic	X	X	X				X	X	X
Poisons, Organic	X	X	X	X	X	X			
Water Reactives	X	X	X	X	X	X			
Organic Solvents	X	X		X	X	X			

7.2 Corrosives (Acids and Bases)

Acids and bases are among the most reactive and corrosive chemicals used in the manufacture of semiconductor components (e.g., etch, strip and clean operations). Acids and bases must always be poured, used, and disposed of with caution by employees who have been trained to work with these chemicals.

- Individual bottles of corrosives should be transported in resistant plastic or rubber buckets. Multiple bottles should be transported on an approved cart with side rails and secondary containment.

- In work areas, individual bottles of corrosives should be stored in secondary containment designed for this purpose. Solvent and oxidizer bottles should not be stored with acid bottles.

- Acid beakers or containers should be filled and used only within an air-exhausted hood. Open-air filling is normally prohibited, and open containers should not be transported.

- Acids or bases should not be mixed with solvents. Corrosive wastes should not be poured into the waste solvent drain or solvent waste containers.

- Water should not be poured into acid. *A*lways *A*dd *A*cid (AAA) to water. This rule also holds true when mixing bases and water.

- Hydrogen peroxide should be poured into acid with thorough stirring after each addition. Note that this differs from the "Always Add Acid" rule for water.

- Acids should be disposed of by flushing with copious amounts of water in an acid sink or with an acid aspirator.

- Never store hydrofluoric acid (HF) in glass or quartz containers. HF can etch the glass container weakening it. HF must be used and stored in resistant plastic containers.

7.3 Solvents (Including Resists)

Organic solvents such as xylene, butyl acetate, and methyl alcohol are used in many work areas. Breathing solvent vapors and skin contact with the liquids must be avoided. An important hazard associated with many of these chemicals is the flammability of the vapors. Solvents should always be treated with respect. Only trained employees should pour, use, or dispose of solvents.

- Individual solvent bottles should be transported in resistant plastic or rubber buckets. Multiple bottles should be transported on an approved cart or wagon with side rails and a drip tray. Large 5-gallon canisters should be transported on a cart or wagon.

- In work areas, individual solvent bottles should be stored in containers specifically designed for that purpose. Solvent bottles should not be mixed with acids, bases, or oxidizers.

- No more than 6 individual 1-gallon bottles of solvent or the chemicals necessary for one shift (whichever is less) should be stored in a work area. Storage in excess of 6 gallons should be in safety-approved flammable liquid storage cabinets or pass-throughs.

- Solvent beakers or containers should only be filled and used within an air-exhausted hood. Open-air filling is normally prohibited and open containers should not be transported.

- Solvents should not be mixed with acids, bases, or oxidizers.

- Solvent waste should not be poured into the waste acid drain, but should be disposed of in a solvent drain or collected in Factory Mutual (FM) approved safety cans.

- Ignition sources such as heat, spark, or open flame should be kept away from solvents.

- Pressurized solvent canisters should be bonded and grounded with an alligator clip and grounding strap to avoid static electricity buildup. Also, they should be placed in secondary containment to contain minor leaks.

- Pressure in solvent canisters should be kept at least 3 psig below the pressure rating of the canister's relief valve. Example: A solvent canister with a pressure relief valve rating of 15 psig should not be pressurized above 12 psig.

- Only nitrogen or other inert gases should be used to pressurize canisters.

- Flammable and combustible solvents should not be stored in plastic bottles. Once the original shipping bottle has been opened, the contents should be transferred to a metal FM approved safety can.

- Paper and other trash that is contaminated with solvents should be disposed of in Factory Mutual-approved flammable solvent waste safety cans.

Check with your supervisor before heating solvents. Heating a solvent can increase its flammability and will increase the amount of solvent vapors given off. Also, heating a chemical will increase subsequent chemical reactions.

Some organic compounds can be absorbed through the skin in significant quantities. Particular care should be taken to avoid skin contact with these chemicals (see Section 5.1). This list includes:

2-Butoxyethanol	2-Ethoxyethanol	Methyl alcohol
2-Butoxyethyl acetate	2-Ethoxyethyl acetate	Phenol
Carbon tetrachloride	Hydrazine	Resists
Dimethyl formamide	2-Methoxyethanol	Toluene
Dioxane	2-Methoxyethyl acetate	Xylene

7.4 Oxidizers

Because oxidizers, such as hydrogen peroxide, react violently, they should not be mixed with solvents. The storage and handling procedures for acids and bases should be followed when working with oxidizers. Oxidizers should be disposed of by aspirating with large amounts of water.

7.5 Process Gases

Process gases represent one of the more severe hazards in the manufacture of electronic components. In particular, some dopants (e.g., arsine, phosphine, diborane) are poisonous and must be used in well-engineered, closed systems.

- Gas cylinders should be disconnected, changed, or moved by employees who are trained to work with these chemicals.

- HPM gases should be confined to source bottle/process equipment that is fully enclosed and exhausted to prevent escape of gas into work areas.

- Lines conveying HPM gases should not have threaded fittings or valves outside exhausted enclosures. All fittings outside of exhausted enclosures should be welded.

- Continuous gas monitoring devices should be provided to detect low-level leakage of toxic gases which have a TLV or PEL below the odor threshold (e.g., arsine, diborane). Toxic gas equipment should not be operated if the monitoring device is not working properly.

- Gas bottles should be secured with a noncombustible safety chain or strap. Cylinder valves should be closed when the cylinder is empty, before it is moved, or if a cylinder contains HPM gas and the process is going to be shut down for an extended period of time (i.e., more than 2 days).

- All cylinders should be capped when the cylinders are not connected for use. Unless cylinders are secured on a special truck or rack, regulators shall be removed and valve caps, when provided for, should be put in place before cylinders are moved.

- Silane and phosphine are pyrophoric chemicals; that is, they can ignite and burn on contact with air. Fire extinguishers are ineffective in controlling fires from such gas leaks. The GAS SUPPLY MUST BE TURNED OFF in these situations. Only employees trained to work with hazardous gases should operate gas supply valves.

- **If an HPM gas leak is suspected, the affected area supervisor or designate should evacuate the area and report the problem over the emergency line. Suspected leaks, whether detected by monitoring devices, odor, or other means, should always be reported. If exposure to dopant or acid gases is suspected, the affected employee(s) must immediately get medical attention.**

- Source bottles and process equipment should be completely leak checked by a qualified technician after a suspected leak has been reported.

NOTE: Diffusion furnace jungle doors should be kept closed and side panels maintained in place at all times. Open doors for preventive maintenance or valve opening/closing should be the only exceptions permitted.

7.6 New Chemicals

Processes are routinely developed that require the use of chemicals new to the facility. These chemicals must be evaluated for unique or unusual hazards so that necessary precautions can be taken.

- A Material Safety Data Sheet should be obtained before a chemical new to the facility is brought on site.

- Employees should receive information and training whenever a new hazard is introduced into their work area.

- The characteristics of specialty chemicals should be evaluated by the safety and environmental departments before the materials are introduced into the work area.

- Any special precautions for specialty chemicals should be enforced by work area supervisors.

8.0 Radiation Safety

8.1 General Radiation Information

Radiation is divided into two major categories: ionizing and nonionizing. Nonionizing radiation refers to the portion of the electromagnetic spectrum, such as visible light and radiofrequency radiation, that is not powerful enough to ionize matter. Ionizing radiation includes x-rays and gamma rays (see Figure 2).

X-rays can come from high-voltage equipment and gamma rays from radioactive material. In addition to gamma rays, radioactive materials can also emit ionizing radiation in the form of small particles such as alpha or beta particles.

When most people see the word *radiation* they think of ionizing radiation and the health hazards associated with atomic bombs and nuclear power plant meltdowns (e.g., death, sterility, leukemia, hair loss). However, most industrial ionizing radiation sources do not generate enough radiation to produce these acute radiation effects.

The major concern with industrial ionizing radiation sources — including those used in the electronics industry — is the potential for the radiation to increase a person's chances of getting cancer. Exposure to radiation does not make a person radioactive nor does it cause a person to glow in the dark.

Nonionizing radiation includes ultraviolet light, visible light, infrared, microwave, radiofrequency, and lasers. Ultraviolet light, microwave, radiofrequency radiation, and lasers are the most hazardous forms used in semiconductor manufacturing.

Radiation exposures can be reduced by:

- Enclosing the radiation source in shielding material
- Staying as far away from the radiation source as practical (not a control for laser beams)
- Limiting the amount of time spent near the source (not a control for laser beams)

Consult your supervisor or safety department for additional information on radiation safety.

8.2 Radioactive Materials

Radioactive material is measured in curies, millicuries, and microcuries (one curie equals 1000 millicuries, one millicurie equals 1000 microcuries). Most radioactive material used in the electronics industry is sealed to ensure the material does not contaminate the work area. One exception to this is krypton-85 gas used in some fine-leak detection systems. A variety of controls are used to keep the gas from leaking into the work area. Check with your supervisor or safety department for safety procedures used for handling radioactive material.

8.3 Radiation Machines

High-voltage equipment above 15,000 volts can produce x-ray radiation. However, at 15,000 volts the x-rays are weak and most of the radiation cannot penetrate the chamber where they are formed (e.g., MEBES [moving electro-beam exposure system], color televisions).

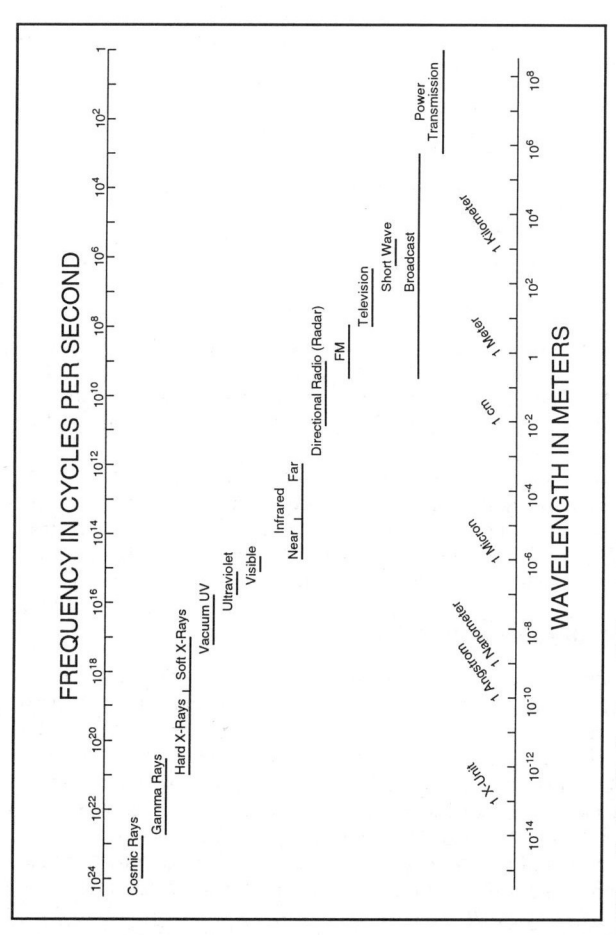

Figure 2. The Electromagnetic Spectrum

Equipment used in integrated circuit (IC) manufacturing that has the potential for creating x-rays includes:

- Cabinet radiography units
- Electron microscopes
- Ion beam milling equipment
- Ion implanters
- X-ray diffraction units
- X-ray fluorescence machines

These units are designed to emit less than 0.5 millirem per hour at the surface of the equipment and are used in uncontrolled radiation areas. They are designed to be operated with no radiation training or radiation film badges. However, local regulations may place restrictions on their use.

Radiation-producing equipment should be checked for radiation leakage when it is installed, modified, or moved to ensure the shielding has not been compromised. Contact your supervisor or Radiation Safety Officer (RSO) if you would like specific information on the ionizing radiation equipment in your work area.

8.4 Ultraviolet (UV) Light

Ultraviolet light can cause acute effects such as reddening of the skin (sunburn) and the feeling of dust in the eyes (welder's flash). Also, overexposure to UV can increase a person's chances of getting skin cancer, especially in fair-skinned and fair-haired individuals.

Under normal operation, most UV sources used in semiconductor manufacturing are less hazardous than exposure to sunlight on a bright day. While the UV sources used in IC production produce a great deal of radiation, most of it is contained in the equipment and is not at the wavelengths associated with adverse health effects.

Filters provided on UV inspection lamps should not be removed. Removing the filters will greatly increase the amount of UV light emitted from the lamps.

8.5 Radiofrequency/Microwave (RF/MW) Radiation

The major health concern with these forms of nonionizing radiation is their heating effects. Both RF and microwaves can heat exposed skin and eyes causing damage. Unlike other sources of heat, some RF/MW frequencies cause heating below the surface of the skin where their effects cannot always be felt. Equipment used in semiconductor and electronic equipment manufacturing that produces RF/MW radiation includes:

- Deposition systems (RF type)
- Microwave ovens
- Plasma ashers
- Plasma descummers
- Plasma etchers
- Radios (when transmitting)
- Rapid thermal anneal
- Resist strippers
- Sputterers

RF/MW exposures are best controlled by enclosing the radiation source in a metal cabinet. Exposures can also be minimized by staying as far away from the radiation source as practical and by limiting the amount of time spent near the RF/MW equipment.

Contact your supervisor or safety department if you would like specific information on the RF/MW equipment in your work area.

8.6 Lasers

The major health hazard of lasers is eye damage and to a lesser extent skin damage. Unlike other forms of radiation, laser beams can travel great distances without a significant loss of power. Therefore, keeping far away from a laser may not provide adequate protection. Lasers are classified according to their potential hazards, with Class 1 lasers being the least hazardous and Class 4 the most hazardous.

When Class 4 lasers are enclosed in a protective housing they can be considered Class 1 lasers except when the protective housing is removed and the interlocks defeated. When a protective housing containing a high-powered laser is removed, specific safety precautions are required such as laser protective goggles or spectacles. ANSI Z136.1-1993, "American National Standard for the Safe Use of Lasers," gives detailed safety precautions for working with lasers.

9.0 Safety and Health Information

9.1 Medical and Exposure Records

Exposure monitoring should be done when it is reasonable to suspect a person may be exposed to a chemical or radiation above one-half the threshold limit value or permissible exposure limit. Medical monitoring is typically done for respirator wearers, persons exposed to chemicals above a certain level, persons working with high-powered lasers, and ERT members.

Employees, former employees, and their designated representatives should have access to exposure and medical records pertaining directly to them or their immediate work environment.

Access should also be provided to members of management having a need-to-know and to certain authorized government representatives. Contact your supervisor, safety department, or medical department if you want copies of your medical or exposure monitoring records.

Employee medical records are records which show the condition of an employee's health. Examples of such records are:

- Medical complaints
- Medical exam results
- Medical opinions, treatments, and diagnoses
- Questionnaires

Employee exposure monitoring records are records which show measurements of a toxic substance or a harmful physical agent. Examples of such records are:

- Biological monitoring records
- Material Safety Data Sheets (MSDSs)
- Workplace monitoring records (e.g., industrial hygiene data)

MSDSs can be requested by contacting your supervisor or safety department. Detailed information on the format of MSDSs is provided in the next section.

9.2 Material Safety Data Sheets

A Material Safety Data Sheet (MSDS) is a document which provides important information and a chemical breakdown of a particular hazardous substance or mixture. An MSDS is compiled by the manufacturer, formulator, distributor, or importer of the substance or mixture and is often required to be shipped along with the substance.

There is no standard format for an MSDS. Figure 3 represents a typical MSDS. While not all MSDSs follow this exact format, they all cover the same topics.

The following information is divided according to the 11 main sections listed in the MSDS used in Figure 3. It outlines information found in an MSDS and contains definitions of key words used in an MSDS. Definitions of key words used in this book are located in Section 2.0.

A. General Information

Trade name; CAS No.; chemical name and/or synonyms; formula; molecular weight; manufacturer's name, address, and telephone number; issue date.

CAS No. The Chemical Abstracts Service registry number for the chemical.

B. First-Aid Measures

Emergency and First-Aid Procedures The procedures for treatment of an individual after overexposure to a specific hazardous substance through inhalation, ingestion, and skin or eye contact. Emergency phone number.

C. Hazards Information

Routes of Exposure The different ways in which the substance can enter the body (e.g., inhalation, ingestion, skin, and eye exposure).

Symptoms of Overexposure The most common sensations or symptoms a person could expect to experience from overexposure to the substance or its components.

Effects or Risks From Exposure The health hazards associated with the substance or its components.

Permissible Concentration in Air The Threshold Limit Value and Permissible Exposure Limit:

Threshold Limit Value (TLV) Threshold limit values refer to airborne concentrations (either short-term or full-shift time-weighted averages) of substances and represent conditions under which it is believed that nearly all workers may be repeatedly exposed day after day without adverse effect.

Permissible Exposure Limit (PEL) The maximum permitted 8-hour time-weighted average concentration of an airborne contaminant established by the U.S. Occupational Safety and Health Administration (OSHA).

Unusual Chronic Toxicity Indicates if the substance or its components are considered potentially cancer-causing by the National Toxicology Program (NTP), the International Agency for Research on Cancer (IARC), or the U.S. Occupational Safety and Health Administration (OSHA).

Flash Point The lowest temperature at which a liquid will give off sufficient flammable vapor for ignition to occur. Values can be determined by open cup (OC) or closed cup (CC) methods.

Autoignition Temperature The minimum temperature required to initiate or cause self-sustained combustion in any substance in the absence of a spark or flame.

Flammable Limits in Air The range of gas or vapor concentrations (percent by volumes in air) that will burn or explode if an ignition source is present. Limiting concentrations are commonly called the lower explosive limits (LEL) and the upper explosive limits (UEL). Below the LEL the mixture is too lean to burn, and above the UEL it is too rich to burn.

Unusual Fire and Explosion Hazards Hazards which may occur as a result of overheating or ignition of the material (including chemical reactions or changes in chemical form or composition).

General Chemical

PRODUCT SAFETY DATA SHEET

ACETIC ACID, GLACIAL

A. GENERAL INFORMATION

TRADE NAME (COMMON NAME)	☒ C.A.S. NO. ☐ GENERAL PRODUCT CODE #
ACETIC ACID, GLACIAL (various grades)	64-19-7 (not limited to glacial acetic acid)

CHEMICAL NAME AND/OR SYNONYM
Acetic Acid, Ethanoic Acid, Methane Carboxylic Acid

FORMULA	MOLECULAR WEIGHT
CH_3COOH	60.05

ADDRESS (No., STREET, CITY, STATE AND ZIP CODE)
GENERAL CHEMICAL CORPORATION

CONTACT	PHONE NUMBER	LAST ISSUE DATE	CURRENT ISSUE DATE
Director Environmental Matters			Sept. 1986

B. FIRST AID MEASURES

EMERGENCY PHONE NUMBER

Call a physician at once

Inhalation: Remove to fresh air. If not breathing, give artificial respiration, preferably mouth-to-mouth. If breathing is difficult, give oxygen if qualified operator is available.

Ingestion: Do not induce vomiting. Give tap water, milk (2-3 glasses), milk of magnesia or egg whites beaten with water. Never give anything by mouth to an unconscious person.

Skin: Immediately flush with plenty of water for 15 minutes, while removing contaminated clothing. Wash clothing before reuse.

Eyes: Immediately flush eyes with plenty of water for at least 15 minutes. If irritation persists, flush an additional 15 minutes.

Seek medical assistance for inhalation, ingestion and irritation.

C. HAZARDS INFORMATION
HEALTH

INHALATION

Vapor causes irritation of eyes, nose and throat; can affect respiratory response, cause coughing and chest pains. TCLo, human = 816 ppm/3M. LC_{50} mouse = 5620 ppm/1 hr.

INGESTION

Severe pain in mouth, gullet, stomach; possible circulatory collapse, uremia and death (human). LD_{50} (oral-rat) = 3310 mg/kg.

SKIN

Concentrated solution causes severe burns if not promptly washed off. Vapor may blacken skin and exposed teeth. Dilute solutions may cause dermatitis in some sensitive individuals. LD_{50} (skin-rabbit) = 1060 mg/kg.

EYES

Causes severe burns. May cause permanent corneal injury, which may be followed by blindness. High vapor concentrations may result in conjunctivitis.

PERMISSIBLE CONCENTRATION AIR (SEE SECTION J) OSHA/TWA: 10 ppm ACGIH/TLV: 10 ppm	BIOLOGICAL None established.

UNUSUAL CHRONIC TOXICITY

Irritation of respiratory tract, chronic bronchitis, erosion of the teeth.

CC124-366 (11/84)   ND - NOT DETERMINED   NA - NOT APPLICABLE

Figure 3. Material Safety Data Sheet

C. HAZARDS (Cont.)

FIRE AND EXPLOSION

FLASH POINT	40 ºC	AUTO IGNITION TEMPERATURE	.426-465 ºC	FLAMMABLE LIMITS IN AIR (% BY VOL.)	
☐ OPEN CUP	☒ CLOSED CUP	(Data sources vary)		LOWER – 5.4	UPPER – 16.0

UNUSUAL FIRE AND EXPLOSION HAZARDS

See Hazardous Decomposition Products, Section G. Gives off flammable vapor above its flash point (104°F). It is dangerous in contact with Chromic Acid, Sodium Peroxide, Nitric Acid or other oxidizing materials.

D. PRECAUTIONS/PROCEDURES

FIRE EXTINGUISHING AGENTS RECOMMENDED

Water spray, foam, CO_2 or dry chemical.

FIRE EXTINGUISHING AGENTS TO AVOID

Water in a solid stream will scatter and spread fire. Use water spray.

SPECIAL FIRE FIGHTING PRECAUTIONS

Use water spray to cool containers exposed to fire and to protect those involved in fire-fighting. Involved personnel should be self-contained breathing apparatus and eye protection and full protective clothing, as needed. Water may be used to dilute spilled material, to reduce flammability and to dissipate irritating vapors evolving from the fire.

VENTILATION

Provide adequate exhaust ventilation to meet TLV/TWA requirements.
<u>Local exhaust</u> — from floor and low spaces. <u>Mechanical (general)</u> — floor-mounted fan or blower. Material should preferably be used in an exhausted hood. Equipment should be explosion-proof and exhaust ducts should be acid-resistant.

NORMAL HANDLING

Avoid breathing vapors. Avoid liquid or vapor contact with eyes, skin or clothing. Use with adequate ventilation. Keep away from heat, sparks and open flame (also, electrical equipment and wiring). Wear personal protective equipment as needed.

STORAGE

Store in closed containers in a well-ventilated area. Outdoors or detached storage is preferred. Keep away from oxidizing agents and combustible materials. Keep above its freezing point (62°F) to avoid rupture of carboys and glass containers due to expansion upon solidification. If frozen, thaw by carefully moving container to warm area.

SPILL OR LEAK (ALWAYS WEAR PERSONAL PROTECTIVE EQUIPMENT – SECTION E)

Provide ventilation. Clean-up personnel should wear self-contained breathing apparatus and personal protective equipment to prevent liquid contact. Use water spray to disperse vapors and protect involved personnel and to reduce flammability. Eliminate ignition sources. Small spills can be neutralized with soda ash or sodium bicarbonate. Large spills should be contained and collected in covered containers, if possible; then labeled for eventual waste disposal.

SPECIAL PRECAUTIONS/PROCEDURES/LABEL INSTRUCTIONS   **SIGNAL WORD – DANGER!**

Water-diluted acid can react with metals to produce hydrogen gas. Label instructions read: Causes severe burns. May be fatal if swallowed. Harmful if inhaled. Combustible. Poison.

E. PERSONAL PROTECTIVE EQUIPMENT

RESPIRATORY PROTECTION

Where vapors are below 500 ppm, use a chemical cartridge, organic vapor respirator with full facepiece or a self-contained breathing apparatus with full facepiece, NIOSH-approved. Vapors to 1000 ppm require an air-supplied respirator with full facepiece, approved by NIOSH. Escape conditions call for NIOSH-approved gas mask with organic vapor canister or self-contained breathing apparatus.

EYES AND FACE

For material handled normally in closed ventilated system, wear safety glasses with nonperforated side shields. For leak, spill or other emergency, wear chemical safety goggles or full face shield to guard against splashing. Do not wear contact lenses under these conditions.

HANDS, ARMS, AND BODY

Wear impervious gloves (rubber or neoprene) and rubber apron. In case of leak, spill or emergency situations with possibility of contact with the material, add full impervious clothing.

OTHER CLOTHING AND EQUIPMENT

Rubber boots, depending on conditions. Eyewash stations and safety showers should be provided in areas of use or handling. Provide protective carriers for handling material in glass bottles.

2

Figure 3. Material Safety Data Sheet (continued)

F. PHYSICAL DATA

MATERIAL IS (AT NORMAL CONDITIONS):	APPEARANCE AND ODOR		
☒ LIQUID ☐ SOLID ☐ GAS	Clear, colorless liquid with a sour, pungent odor resembling vinegar.		
BOILING POINT 117.9 °C	SPECIFIC GRAVITY (H_2O = 1) (liquid) 1.05		VAPOR DENSITY (AIR = 1) 2.07
MELTING POINT 16.7 °C			
SOLUBILITY IN WATER (% by Weight) Complete	pH (Aqueous Solutions) 2.4 @ 1.0M 2.9 @ 0.1M 3.4 @ 0.01M		VAPOR PRESSURE (mm Hg at 20°C) ☒ (PSIG) ☐ 11.4
EVAPORATION RATE (Butyl Acetate = 1) ☐ (Ether = 1) ☒ 11.0 (Butyl Acetate = 1): 0.97	% VOLATILES BY VOLUME (At 20°C) 100		

G. REACTIVITY DATA

STABILITY	CONDITIONS TO AVOID
☐ UNSTABLE ☒ STABLE	None.

INCOMPATIBILITY (MATERIALS TO AVOID)

Strong oxidizing agents, hydroxides, amines, oxides, carbonates. Avoid contamination with acetaldehyde, chromic acid and alkalies.

HAZARDOUS DECOMPOSITION PRODUCTS

We would expect burning to produce carbon monoxide and/or carbon dioxide.

HAZARDOUS POLYMERIZATION	CONDITIONS TO AVOID
☐ MAY OCCUR ☒ WILL NOT OCCUR	None.

H. HAZARDOUS INGREDIENTS (Mixtures Only)

MATERIAL OR COMPONENT/C.A.S. #	WT. %	HAZARD DATA (SEE SECT. J)
Not Applicable.		

CC124-366 (11/84)   3   * = PROPRIETARY – TRADE SECRET

Figure 3. Material Safety Data Sheet (continued)

I. ENVIRONMENTAL

DEGRADABILITY/AQUATIC TOXICITY	OCTANOL/WATER PARTITION COEFFICIENT
BOD_5 (g/gl): 0.34-0.88 Std. Dilution/Sewage Seed	−0.17

Aquatic Toxicity Rating: TLm96: 100-10 ppm [TLm96 = Lethal Concentration (50% kill) 96 days]

EPA HAZARDOUS SUBSTANCE? (CLEAN WATER ACT SECT. 311) ☒ YES ☐ NO	IF SO, REPORTABLE QUANTITY 5000 #	40 CFR 116-117

WASTE DISPOSAL METHODS (DISPOSER MUST COMPLY WITH FEDERAL, STATE AND LOCAL DISPOSAL OR DISCHARGE LAWS)

Treatment or disposal of waste generated by use of this product should be reviewed in terms of applicable federal, state and local laws and regulations. Users are advised to consult with appropriate regulatory agencies before discharging, treatment or disposing of this material. Where permitted under above-mentioned regulations, waste material can be incinerated in a furnace (supplementary fuel may be necessary for burning); or neutralized waste disposed of in an approved chemical wastes landfill.

RCRA STATUS OF UNUSED MATERIAL IF DISCARDED	HAZARDOUS WASTE NUMBER (IF APPLICABLE)	40 CFR 261
Not a "hazardous waste".	N.A.	

J. REFERENCES

PERMISSIBLE CONCENTRATION REFERENCES:

TWA: OSHA Regulations, 29CFR 1910 (1982), "Z List"
TLV: ACGIH 1984-85 List, "Threshold Limit Values for Chemical Substances ...",
NIOSH Registry (RTECS), 1981-82, Accession No. AF1225000, "Acetic Acid".

REGULATORY STANDARDS	D.O.T. CLASSIFICATION Corrosive Material	49 CFR 173
D.O.T. Hazardous Materials Table 49CFR 172.101	I.D. No.: UN2789	

GENERAL:

Merck Index, 10th ed., 1983
Verschueren: "Handbook of Environmental Data on Organic Chemicals", 2nd ed., 1983, Van Nostrand Reinhold.
U.S. Coast Guard: CHRIS Manual, Entry: "Acetic Acid"
Technical Guide No. 6, "Handbook of Organic Industrial Solvents", 5th ed., 1980, Alliance of American Insurers, 20 N. Wacker Drive, Chicago, IL 60606.
Emergency Action Guides, Association of American Railroads, Washington, DC (1984).
Fire Hazards: NFPA Manual 49, "Hazardous Chemicals Data", 8th ed. (1984).

K. ADDITIONAL INFORMATION

For Manufacturing Use Only.

Not For Food Or Drug Use.

PSDS FILE #- GC 4003

THIS PRODUCT SAFETY DATA SHEET IS OFFERED SOLELY FOR YOUR INFORMATION, CONSIDERATION AND INVESTIGATION.

GENERAL CHEMICAL CORPORATION PROVIDES NO WARRANTIES, EITHER EXPRESS OR IMPLIED, AND ASSUMES NO RESPONSIBILITY FOR THE ACCURACY OR COMPLETENESS OF THE DATA CONTAINED HEREIN.

Figure 3. Material Safety Data Sheet (continued)

NFPA Hazard Classes Hazard identification system established by the National Fire Protection Association which gives a general idea of the inherent hazards of the chemical and the order of the severity of the hazards under emergency conditions such as spills, leaks, and fires. This system identifies the "flammability" (red), "health" (blue), and "reactivity" (yellow) of a chemical and indicates the order of severity of each hazard by use of one of five numerical gradings, from four (4), indicating the severe hazard or extreme danger, to zero (0), indicating no special hazard.

D. Precautions/Procedures

Fire-Extinguishing Agents Recommended Firefighting substances determined to be suitable for use on the burning material.

Fire-Extinguishing Agents to Avoid Firefighting substances and/or methods which might be hazardous.

Special Firefighting Precautions Requirements concerning use of specialized fire-extinguishing materials or techniques.

Ventilation The removal or addition of air to minimize the potential hazards (fire, inhalation, irritation) associated with the substance.

Normal Handling General precautionary measures to be taken when handling the substance.

Storage General precautionary measures to be taken when handling and storing the substance.

Spill or Leak Procedures The methods to be used to control and clean up spills and leaks.

Special Precautions/Procedures/Label Instructions
Unique hazards and controls for the chemical; wording recommended by the manufacturer for chemical warning labels.

E. Personal Protective Equipment

Respiratory Protection Specific type of respiratory equipment recommended while handling the substance outside of an air-exhausted work station or a closed system.

Eyes and Face The specific type of eye protection recommended.

Hands, Arms, and Body Type of gloves, apron, or other body covering recommended to provide proper protection from skin contact.

Other Clothing and Equipment Additional equipment needed for protection against possible overexposure.

F. Physical Data

Material Is (at Normal Conditions) Physical state the chemical is at room temperature (e.g., solid, liquid, or gas).

Appearance and Odor The physical coloration; the characteristic odor of a substance and odor threshold (the lowest concentration at which a chemical will have an odor).

Boiling Point The temperature at which a liquid changes to the vapor state. Usually stated in degrees Fahrenheit (°F) or degrees centigrade (°C) at sea level pressure — 760 millimeters (mm) of mercury (Hg). For mixtures, the initial boiling point of the boiling range may be given.

Melting Point The temperature at which a substance's solid phase is in equilibrium with the liquid phase at room temperature.

Specific Gravity The ratio of the weight of a material to the weight of an equal volume of water. In the case of immiscible liquids, the specific gravity will predict whether the substance will sink or float on water. For example, if the specific gravity is greater than 1, it will sink; if the specific gravity is less than 1, it will float.

Vapor Density The specific gravity of a gas. It equals the ratio of the weight of a vapor or gas (with no air present) compared to the weight of an equal volume of air at the same temperature and pressure. Values greater than 1 indicate that it tends to settle. The effects of temperature and air movement must also be considered. In a clean room the vertical laminar flow (VLF) hoods will normally have a greater effect on the movement of a gas or vapor than will the vapor density.

Solubility in Water The amount of substance which will dissolve in a given amount of water (e.g., 2 g/100 ml).

pH pH is a value used to represent the strength of an acid or base. pH values of 1 to 6 indicate acids, with 1 being a very strong acid and 6 being a very weak acid. Neutral materials, such as pure water, have a pH of 7. Bases have pH values of 8 to 13, with 8 being a very weak base and 13 being a very strong base.

Vapor Pressure The pressure exerted by a saturated vapor above its own liquid in a closed container, usually stated in millimeters (mm) of mercury (Hg) at 68° Fahrenheit (°F) or 20 degrees centigrade (°C).

Evaporation Rate The rate at which a substance will vaporize (evaporate) when compared to the vaporization of a known material, usually butyl acetate.

Percentage Volatile by Volume The percentage by volume of a substance that can evaporate.

G. Reactivity Data

Stability Chemical reactivity of a substance under a given set of conditions.

Conditions to Avoid Self-explanatory.

Incompatibility Chemicals to be avoided both in storage and in use.

Hazardous Decomposition Products Hazardous substances that may be produced as the substance is exposed to burning, oxidation, heating, or reaction with other chemicals.

Hazardous Polymerization Uncontrolled polymerization reaction which results in the release of potentially hazardous amounts of heat or gases.

H. Hazardous Ingredients (Mixtures Only)

A substance in a mixture in sufficient concentration to produce a flammable vapor or gas or to produce acute or chronic adverse effects in persons exposed to the product either during normal use or predictable misuse.

Percentage(s) % The approximate percentage, by weight or volume, that each hazardous substance represents. Health hazards must be listed if the substance comprises 1% or more of the mixture. Carcinogens are listed if they are at least 0.1% of the mixture.

I. Environmental

Degradability/Aquatic Toxicity How fast the chemical breaks down and becomes a hazard to fresh and salt water environments.

EPA Hazardous Substance Indicates whether the chemical is considered a hazardous substance and, if so, when a spill of the material must be reported to the EPA or Coast Guard under 40 CFR 116–117.

Waste Disposal Method Manufacturer's recommended disposal method.

RCRA Status of Unused Material If Discarded Indicates whether the chemical is considered a hazardous waste under Federal Regulations 40 CFR 261 (Resource Conservation and Recovery Act) and its hazardous waste number.

J. References

Permissible Concentration References References used for determining allowable limits for the chemical.

Regulatory Standards Department of Transportation (DOT) classification and regulations covering the material.

General General references used for the information provided in the MSDS.

K. Additional Information

Miscellaneous safety information that does not fit into another category.

9.3 Additional Information

If additional information is needed on any chemical, contact your supervisor or the facility's safety department. Information on the disposal of chemicals can be obtained by contacting the facility's environmental coordinator.

This book is intended to serve only as a guide for the handling of chemicals used in the semiconductor/electronics industry. It is not a substitute for chemical training.

10.0 Trade-Name Compounds and Chemical Mixtures

This section lists generic components of trade-named compounds used in semiconductor manufacturing. This information is based on MSDSs and analytical information provided by an independent laboratory. The individual components of a mixture are listed in order of concentration with the component of largest concentration listed first.

Many MSDSs list only the components that are considered hazardous and therefore do not indicate the complete composition of the mixture. Also, the information obtained by laboratory analysis includes only volatile organic compounds; inorganic and nonvolatile organic compounds may also be present. Changes in formulation can also result in different ingredients for mixtures and trade-name chemicals.

If you need specific information on a compound, consult the MSDS and contact your safety department. This section does not provide the detailed information necessary to use the listed compounds safely.

10.1 Photoresists

Photoresists typically contain three types of compounds:

- Photoreactive
- Resin (positive resists) or polymer (negative resists)
- Solvent

The resins/polymers and the photoreactive compounds found in photoresists evaporate slowly or not at all. Therefore, they are not an inhalation hazard under normal conditions. The solvent portion of a resist makes up about 65% of positive resists and about 90% of negative resists and is normally the major concern. Ingredients for the resists listed in Section 10.2 show only the solvent components. If you want detailed information on a resist, contact your safety department.

10.2 Generic Components

Note: (+) denotes positive resists; (−) denotes negative resists.

Name	Components
1030 Acid Cleaner	Sulfuric acid Formic acid
A20/A30/A40	Tetrachloroethylene Dichlorobenzene Phenol
Accuglass 108/203	Substituted siloxane polymer
Accuspin ASX-10 Spin-On Dopant (also AS-120 and AS-217)	Silicate ester Arsenic Alcohol
Accuspin Spin-On Dopant	Boron trioxide 2-Methoxyethanol
Accustrip NP-200/NP-240U	Sulfonic acid Hydrocarbons
Accustrip P-300	M-Pyrol Substituted piperazines
Agitene 141/Super	Paraffin/naphthalene (blend)
Alcatel 100 Pump Oil	Aliphatic hydrocarbons

Name	Components
Alloy 101/102/110/151/155/194	Copper
Alloy 195	Copper Cobalt
Alloy 210/220/230/240/260	Copper Zinc
Alloy 510/511/521	Copper Tin
Alloy 638	Copper
Alloy 725	Copper Nickel Tin
Alloy 732/735/762/770	Copper Nickel Zinc
Allros 100 Flux	Alcohol White gum rosin
Alpha 100 Flux	Isopropyl alcohol
Alpha 611F Flux	Solvent-alcohol blend White gum rosin Organic activator
Alpha 850-33 Flux	Isopropyl alcohol Organic acid Amine salt
Alpha 994 Cleaner	Water Hydrochloric acid Inorganic acids
Aluminum Etch 16-1-1-2	Phosphoric acid Acetic acid Nitric acid

Name	Components
Aluminum Etch 82-3-15-0	Phosphoric acid Acetic acid Nitric acid
AP 10 Processor	Hydrochloric acid Ferrous chloride
Aqua Regia	Hydrochloric acid Nitric acid
Aqua-Sol Flux	Isopropyl alcohol Nonvolatile organic acids
Araldite ECN 1280	Epoxy cresol novolac resin
Arc Cleaner	Acetone M-Pyrol Hydrocarbons
Arcosolv PM	Propylene glycol methyl ether
Aristoline (+)	2-Ethoxyethyl acetate Butyl acetate Xylene Ethyl benzene
Arklone	Trichlorotrifluoroethane (Freon TF) Water Surfactant
Arklone P	Trichlorotrifluoroethane (Freon TF)
Aron Alpha	Ethyl cyanoacrylate
Arsenosilica Film 0308	Arsenic pentoxide Ethyl acetate Ethyl alcohol Methyl alcohol

Name	Components
As-1/As-1CE/As-5CE/As-18CZ5E/ As-18CZ6E/As-18CZ10A/ As-1400	Ethyl alcohol Ethyl acetate Acetic acid Arsenosilicate polymer
Atomex	Cyanide salts
AZ 303 A Developer	Sodium hydroxide
AZ 312 MIF Developer	Tetramethyl ammonium hydroxide
AZ 351 Developer	Sodium hydroxide
AZ 400K Developer	Potassium hydroxide
AZ 427 MIF Developer	Tetramethyl ammonium hydroxide
AZ 1310-SF (+)	2-Ethoxyethyl acetate Xylene n-Butyl acetate
AZ 1312-SFD (+)	2-Ethoxyethyl acetate n-Butyl acetate Xylene
AZ 1318-SFD (+)	2-Ethoxyethyl acetate n-Butyl acetate Xylene
AZ 1350J (+)	2-Ethoxyethyl acetate n-Butyl acetate Xylene
AZ 1370 (+)	2-Ethoxyethyl acetate Xylene n-Butyl acetate
AZ 1370-SF (+)	2-Ethoxyethyl acetate Xylene n-Butyl acetate

Name	Components
AZ 1375 (+)	2-Ethoxyethyl acetate Xylene n-Butyl acetate
AZ 1470 (+)	2-Ethoxyethyl acetate Xylene n-Butyl acetate Ethyl benzene
AZ 1512 (+)	Propylene glycol monomethyl ether acetate
AZ 1512-SFD (+)	Propylene glycol monomethyl ether acetate
AZ 1518 (+)	Propylene glycol monomethyl ether acetate
AZ 1518-SFD (+)	Propylene glycol monomethyl ether acetate
AZ 4140 (+)	2-Ethoxyethyl acetate n-Butyl acetate Xylene
AZ 4210 (+)	2-Ethoxyethyl acetate Xylene n-Butyl acetate Ethyl benzene
AZ 4330 (+)	2-Ethoxyethyl acetate Xylene n-Butyl acetate
AZ 4620 (+)	2-Ethoxyethyl acetate n-Butyl acetate Xylene
AZ 5214	1-Methoxy-2-propyl acetate
AZ Developer	Dilute aqueous solution of alkaline salts, principally phosphates

Name	Components
AZ EMB 70/30 AZ EBR	Propylene glycol monomethyl ether Propylene glycol monomethyl ether acetate
AZ Protective Coating	2-Ethoxyethyl acetate
AZ Thinner	2-Ethoxyethyl acetate n-Butyl acetate Xylene
B239	Sodium nitrite
B446	Ethyl acetate Ethyl alcohol
B-Etch	Ammonium fluoride Hydrofluoric acid
Backlap Slurry	D.I. water Aluminum oxide Ethanolamine Sodium chromate Diethanolamine
Baker PRS-1000R, Positive Resist Stripper	M-Pyrol Sulfolane 2-(2-Ethoxyethoxy) ethanol Tetraethylene glycol Monoethanolamine
Binary Doped Silane/Phosphine Mix	Phosphine Silane
Black Marking Ink, 105E Multilayer Dielectric Ink HD 2011	Acrylated epoxy Lead borosilicate Lead compounds (insoluble)
Blanco-Tron TF	Trichlorotrifluoroethane (Freon TF)
Blanco-Tron TMC	Trichlorotrifluoroethane (Freon TF) Methylene chloride

Name	Components
Blanco-Tron TMS Plus	Trichlorotrifluoroethane (Freon TF) Ethyl alcohol Methyl alcohol Nitromethane
Blue Epoxy Ink	Diacetone alcohol Blue dye
BlueShield 1/2/3	Inert gases
BlueShield 6/7/8	Inert gases Carbon dioxide
Bluestone Blue Vitrol	Copper sulfate
BOE (Buffered Oxide Etch)	Ammonium fluoride Hydrofluoric acid
Boron B-30/B-40/B-50/B-60	Ethyl alcohol Ethyl acetate Glycerin Isopropyl alcohol Methyl alcohol Boric acid Borosilicate polymer
Boydes PTS Developer	Potassium hydroxide Hydroquinone Bromide salts
Bravo Extra Heavy Duty Stripper	Water Alkali metasilicate Ethanolamine Sodium hydroxide
Brightener E-3	Arsenic
Burmar 712D	1,2,4-Trichlorobenzene Dodecylbenzenesulfonic acid Alkyl benzenes Phenol

Name	Components
Burmar Lab Clean	Ammonium hydroxide Isopropyl alcohol M-Pyrol
Burmar Nophenol-922 HB	Alkyl benzenes Dodecylbenzenesulfonic acid Ethylene glycol phenyl ether Catechol
Burmar Nophenol-944	Dodecylbenzenesulfonic acid Alkyl benzenes Ethylene glycol phenyl ether
Burmar Posistrip 830	M-Pyrol 2-(2-Aminoethoxy) ethanol
Burmar Posistrip LE	Morpholine M-Pyrol Butyrolactone
Burmar SA-80	Ammonium persulfate
C-521	Sodium nitrite
C-585	Potassium hydroxide
C-589	Isopropyl alcohol
C-596	Maleic acid polymers
C-P 8 Solution	Nitric acid Hydrofluoric acid
Carro's Acid	Sulfuric acid Hydrogen peroxide
Castolite-AC	Styrene monomer
Castolite Hardener 20-8122/20-8126	Dimethyl phthalate Methyl ethyl ketone peroxide
Catalyst A	2-Butoxy ethanol

Name	Components
CB-15 Clearing Bath	Alkaline corrosive
CB-16 Clearing Bath	Sodium hydrogen sulfite
CC-200 Positive Developer	Alkaline (corrosive)
CEA-100 Micro-Chrome Etchant	Ceric ammonium nitrate Acetic acid
Ceilcrete 6650	Styrene
Cem 388	Ethyl benzene Proprietary components
Cem 420	Ethyl benzene n-Butyl alcohol Proprietary components
Cemsa 15	Toluene Methyl phenyl ether
CEN-300 Micro-Chrome Etchant	Nitric acid
Chemtranic Flux Off	Methylene chloride Trichlorotrifluoroethane (Freon TF) Carbon dioxide
Chemtranic Flux Strippers	Methylene chloride Isopropyl alcohol
Chlorothene SM Solvent	1,1,1-Trichloroethane 1,4-Dioxane 1,2-Butylene oxide Nitromethane
Chlorothene VG Solvent	1,1,1-Trichloroethane
Chrome Etch KTI	Water Ceric ammonium nitrate Nitric acid
Chromic #5	Sulfuric acid Chromic acid

Name	Components
Cleaner, Ink Independent	Ethyl acetate Other solvents
Cleaner 944	Hydrochloric acid
Clearview Glass Cleaner	2-Butoxyethanol Ammonia Butane/propane blend
Colec 302	Potassium hydroxide Sodium hydroxide
Contrad 70 Cleaner	Detergent
Copper, Brass Brite Dip 127/1127	Phosphoric acid Acetic acid Nitric acid
Copper 1 Reagent	Water Sodium acetate Acetic acid
Copper 2 Reagent	Water Isopropyl alcohol
Copperlite RD-25	Sulfuric acid Inorganic acids Nitric acid
CP290B Activator	Trisodium phosphate Glycerin Methylaminoethanol Sodium sulfite Isopropyl alcohol
Cr-9 Chromium Etchant	Perchloric acid
CR-9 Etch	Perchloric acid

Name	Components
Cronaflex PDC Developer	Sodium potassium sulfite Hydroquinone Sodium gluconate Potassium carbonate Potassium hydroxide
Cronaflex PFC Developer	Ammonium thiosulfate Sodium bisulfite Acetic acid tetraacetate Trisodium hydrogen ethylene diamine
CuBath MD Addition	Sulfuric acid
CuBath M-HY	Sulfuric acid Organic sulfide
CuBath M Make-up	Sulfuric acid Copper sulfate
Cuposit CP-74A Electroless Copper	Dissolved copper Mercuric acetate
Cuposit CP-74B Electroless Copper	Sodium hydroxide
Cuposit CP-74M Electroless Copper	Formaldehyde Dissolved copper
Cuposit CP-74R Copper Replenisher	Formaldehyde Dissolved copper
Cymel Molding Compound	Melamine-formaldehyde resin
6200/6500 Drum Cleaning Solvent	1,1,1-Trichloroethane Kerosene 1,4-Dioxane Nitromethane
DAG 154	Isopropyl alcohol Butyl alcohol 2-Ethoxyethanol

Name	Components
Dazzlens Cleaner (Lens Cleaner M6015)	Isopropyl alcohol Acetic acid Acetone
Deltastrip 70NS	Diethylene glycol monobutyl ether Ethanolamine
Developer 1002	2-Ethoxyethanol
Developer TMD-250	Tetramethyl ammonium hydroxide
Dialec 32	1,1,1-Trichloroethane
DN-10 E Beam Negative Resist Developer	Methyl ethyl ketone Ethyl alcohol
Doped Poly Etch	Nitric acid D.I. water Hydrofluoric acid
Dow Corning 20 Release Coating	Stoddard solvent
Dow Corning 200 Fluid	Dimethyl siloxane
Dow Corning 705 Diffusion Pump Fluid	Silicone
Dow Corning 1200 Prime Coat	Naphtha
DP-13 E Beam Positive Resist Developer	Methyl isoamyl ketone Methyl isopropyl ketone
DP-18 E Beam Positive Resist Developer	Methyl isoamyl ketone Methyl isopropyl ketone
Drierite	Calcium sulfate
Dynalith EPR-5000 (+)	Propylene glycol monomethyl ether acetate
Dynalith OFPR-800PG (+)	Propylene glycol monomethyl ether acetate
Dynasol-3	Alkyl benzenes

Name	Components
Dynasolve 100	Dimethyl formamide
Dynasolve 190	Methyl alcohol Potassium hydroxide
Dynasolve MP-500	2-Ethoxyethanol Potassium hydroxide
Dynasolve MP-710/MP-750	Propyleneglycol monomethyl ether Methyl alcohol Potassium hydroxide
Dynasolve MP Aluminum Grade	2-Ethoxyethanol Potassium hydroxide
6-6 Epoxy Chem Resin Finish Clear Curing Agent	Epoxy resin solids n-Butyl acetate Propylene glycol monomethyl ether Toluene Diethylene glycol monobutyl ether Butyl alcohol
777 Etch	Ammonium fluoride Acetic acid Water Ethylene glycol
Electrocleaner 111	Sodium hydroxide
EME-X8778	Silica Carbon black
EMT-130 Positive Photoresist Stripper	M-Pyrol
EMT-500	M-Pyrol
Enplate Activator 442	Hydrochloric acid Tin

Name	Components
Enplate Conditioner 470	Sulfuric acid Phosphoric acid Chromic acid
Enplate NI-416M/NI-418A	Water Nickel sulfate
Enplate NI-418B	Water Sodium hypophosphite Ammonium hydroxide
Enplate NI-418C	Water Sodium hypophosphite
Enstrip Au-78 (Gold Etch)	Potassium cyanide Lead oxide Proprietary ingredients
Enstrip NP-1	Ammonium hydroxide
Enstrip NP-2	Ethylene diamine sodium diethyldithiocarbamate
Entec 234 (Oil Based Anti-Foam)	Mineral oil Tall oil fatty acids
Entec 309 (Corrosion Inhibitor)	Methyl 1-H-benzoltriazole (1-hydroxyethylidine) bis-phosphoric acid
Entec 310A (Corrosion Inhibitor/ Dispersant)	Methyl 1-H-benzoltriazole Phosphoric acid Potassium hydroxide
Entec 313 (Corrosion Inhibitor/ Dispersant)	Methyl 1-H-benzoltriazole Phosphoric acid Sodium hydroxide Trade secret ingredient

Name	Components
Entec 318 (Corrosion Inhibitor/ Dispersant)	Methyl 1-H-benzoltriazole Phosphoric acid Sodium hydroxide Sodium molybdate Trade secret ingredient
Entec 321 (Dispersant)	No hazardous components listed
Entec 327 (Surfactant)	Ethyl alcohol Isopropyl alcohol
Entec 336 (Corrosion Inhibitor/ Dispersant)	Sodium molybdate Sodium nitrite
Entec 337 (Corrosion Inhibitor)	Sodium bicarbonate
Entec 338 (Corrosion Inhibitor)	Tetrasodium borate, pentahydrate Sodium nitrite
Entec 340	n-Alkyl dimethyl benzyl ammonium chloride Tin (organic)
Entec 343	1,3-Dichlorodimethyl hydantoin Inerts
Entec 345A (Biocide)	2,2-Dibromo-3-nitrilopropionamide
Entec 346 (Biocide)	Sodium dichloro-S-triazinetrione
Entec 349 (Biocide)	Dodecylguanidine hydrochloride Isopropyl alcohol Methylene bis(thiocyanate)
Entec 353 (Corrosion Inhibitor)	Methyl 1-H-benzoltriazole Phosphoric acid Potassium hydroxide
Entec 356 (Corrosion Inhibitor/ Dispersant)	Methyl 1-H-benzoltriazole Phosphoric acid Potassium hydroxide

Name	Components
Entec 363	1-Bromo-3-chloro-5,5-dimethylhydantoin
Entec 364 (Biocide)	1-Bromo-3-chloro-5,5-dimethylhydantoin
Entec 367 (Biocide)	1-Bromo-3-chloro-5,5-dimethylhydantoin
Entec 436 (Corrosion Inhibitor/Dispersant)	Phosphoric acid Sodium molybdate
Entec 763 (Fuel Oil Additive)	Dodecylbenzene sulfonic acid Heavy aromatic naphta 2-Methoxyethanol Sulfuric acid
Entec CPS-4 (Internal Boiler Treatment)	Ethylenediamine tetraacetic acid, tetra sodium salt Sodium hydroxide
Entec Drain Pan Biocide	n-Alkyl dimethyl 1-naphthylmethyl ammonium chloride
EPD 1240 HRP Positive Developer	Alkaline and organic salts
EPF B20 Fixer	Acetic acid Sulfuric acid
Epoxy Chromate Metal Primer	Zinc chromate Methyl ethyl ketone Epoxy resin solids MICA Propylene glycol monomethyl ether Organophilic clay Toluene
Epoxy Cure Agent	Isopropyl alcohol Polyamide resin solids Toluene Organophilic clay

Name	Components
Epoxy Solvent Cure Agent	Methyl ethyl ketone Toluene Butyl alcohol
Everest Top Layer Silver/ PD Ink HC 4305	Silver Lead borosilicate Zinc oxide
4282 Flux	Ethyl alcohol Isopropyl alcohol Hydrochloric acid Methyl alcohol
Fast Cure 45 Epoxy	Aluminum oxide Epoxy resin Polymercaptan Aromatic plasticizer Substituted aminophenol Titanium dioxide Styrene Ethylene glycol Silica
FC-40, FC-77, FC-84, FC-5312	Inert fluorochemicals that are practically nontoxic
FC-95	Isopropyl alcohol Surfactants
Fidelex 218 Metal Stripper	Sodium cyanide Sodium hydroxide
Fidelity 1215A Solder Stripper	Fluoroboric acid Formic acid
Fidelity 1215B Solder Stripper	Fluoroboric acid Formic acid Propylene glycol
Field Test Reagent (Betz Entec)	Sulfuric acid

Name	Components
Filling Solution	Potassium chloride Silver chloride
Film Remover	Isopropyl alcohol
Flakeline 22HT	Styrene
Flocon Antiscalant 100	Water Polyacrylic acid
Fluro-Chek WP-170 Penetrant	Hexylene glycol
Fomblin Y 06/6	Perfluoropolyether
Fomblin Y 18/8 H Vac	Perfluorinated polyethers
Fomblin Y 25/5	Perfluorinated polyethers
Fomblin Y 25/6 L Vac	Perfluorinated polyethers
Fomblin Y 140/13 H Vac	Perfluorinated polyethers
Forming Gas	Nitrogen Hydrogen
Formula 409 Cleaner	2-Butoxyethanol
Freckle Etch	Phosphoric acid Acetic acid Nitric acid Hydrofluoric acid Boric acid
Fremont 5015M Photoresist Remover	Formic acid
Freon-11	Trichlorofluoromethane
Freon-12	Dichlorodifluoromethane
Freon-13	Chlorotrifluoromethane
Freon-13B	Bromotrifluoromethane

Name	Components
Freon-14	Carbon tetrafluoride
Freon-22	Chlorodifluoromethane
Freon-23	Trifluoromethane
Freon-113	Trichlorotrifluoroethane
Freon-116	Hexafluoroethane
Freon MF	Trichlorofluoromethane
Freon TF	Trichlorotrifluoroethane
Freon TMC	Trichlorotrifluoroethane Methylene chloride
Freon TMS	Trichlorotrifluoroethane Methanol Nitromethane
Fuser Lubricant, Silicone Fuser Oil	Polydimethylsiloxane
G-182B Activator	Sodium hydroxide
Genesolv 404 Azeotrope	Trichlorofluoromethane Acetone n-Hexane
Genesolv A	Trichlorotrifluoroethane (Freon TF)
Genesolv B	Trichlorotrifluoroethane (Freon TF)
Genesolv DFX	Trichlorotrifluoroethane (Freon TF) Methyl alcohol Hexane (isohexane) Acetone Nitromethane
Genesolv DM Azeotrope	Trichlorotrifluoroethane (Freon TF) Methylene chloride

Name	Components
Genesolv D Solvent (Various Grades)	Trichlorotrifluoroethane (Freon TF)
Genetron 11	Trichlorofluoromethane (Freon MF)
Glass Cleaner (True Blue)	Isopropyl alcohol 2-Butoxyethanol
Glass Etch	Acetic acid Ammonium fluoride Water
Glid-Guard Epoxy — Safety Blue	Polyamide resin Magnesium silicate (talc) Isopropyl alcohol Methyl ethyl ketone Titanium dioxide Toluene Silica Mineral spirits
Glid-Guard High Solids Epoxy No. 5433	Magnesium silicate, hydrate Polymeric fatty acid amide Toluene 2-Butone 2-Propanol
Glid-Guard High Solids Epoxy No. 5434	Epoxy resin Xylene Amorphous silica Oxohexyl acetate
Gold Au78 Stripper	Potassium cyanide Lead oxide Proprietary ingredients
Gold Cyanide Salt E-59	Cyanide salts in solution
Gold/Platinum Ink A3388	Platinum Lead borosilicate Dibutyl phthylate

Name	Components
Gold/Platinum Ink A3395	Platinum Lead borosilicate Urea
Gold Strip 123	Sodium cyanide
Gold Technistrip	Potassium cyanide Lead oxide
Goodrite NR-D	Stoddard solvent
Goodrite NR-R	n-Butyl acetate
Halocarbon 11	Trichlorofluoromethane
Halocarbon 12	Dichlorodifluoromethane
Halocarbon 14	Tetrafluoromethane
Halocarbon 23	Trifluoromethane
Halocarbon 115	Chloropentafluoroethane
Halocarbon 116	Hexafluoroethane
Hardness 2 Test Solution	Propylene glycol Hydroxylamine hydrochloride Isopropyl alcohol
HCP 600	Potassium pyrophosphate
HCP 825	Polyphosphoric acid
HD Detergent	2-Butoxyethanol Sodium metasilicate
High Grade 1086	Isopropyl alcohol
High Speed Endura-Etch Starter	Copper chloride Ammonia
HI-P Contact Cleaner	Trichlorotrifluoroethane (Freon TF)
HI Sol 15	Alkyl benzenes

Name	Components
H.M.D.S. III	Hexamethyldisilazane 2-Ethoxyethyl acetate
Hollis 600 Tinning Fluid	Polyglycol ether (water soluble)
Humiseal 503/1B31	Toluene Methyl ethyl ketone
Hydranal Composite 5	Diethylene glycol monomethyl ether Imidazole Sulfur dioxide Iodine
Hydrion Buffer Salts pH Value 2.0–4.0	Potassium biphthalate
Hydrion Buffer Salts pH Value 5.0–8.0	Sodium phosphate, dibasic
Hydrion Buffer Salts pH Value 8.0–11.0	Sodium borate, dibasic
Hydrion Buffer Salts pH Value 9.0–11.0	Sodium carbonate
Hydrion pH Buffer Salts pH Value 5.0–8.0	Potassium phosphate, monobasic
Hysol EE-4143	Butyl glycidyl ether
Hysol Hardener HD-3475	Diethylenetriamine
Hysol K0109	Silver Vinyl cyclohexene dioxide
11-1193 Isocut Cutting Fluid	Kerosene Natural oil
IBM Ink	Dye Water Oil (water soluble) Diethylene glycol M-Pyrol

Name	Components
IC 1200 (09)	Boron tribromide
Implanter Fumer	Hydrogen fluoride D.I. water
Ink Remover (Jasco)	Methylene chloride Methyl alcohol
Inland 19 (Varian GP Type Fluid)	Neutral paraffinic oil
Inland 77	Paraffinic hydrocarbons
Inland 80	Aliphatic hydrocarbons
Instafill Component A	Methylene bisphenyl isocyanate High molecular weight oligomers of MDI
Instafill Component B	Polyurethane resin Amine catalyst
Instapak 40 Component A	Methylene bisphenyl isocyanate High molecular weight polymers
Instapak 40 Component B	Trichlorofluoromethane (Freon)
Instapak Holster Solvent	Tripropylene glycol methyl ether
Instapak Port Cleaner	Butyrolactone
Invoil 20	Paraffinic petroleum oil
J100	Tetrachloroethylene Dichlorobenzene Phenol Dodecylbenzene sulfonic acid
JMI Sloop	Tin Lead Silver Rosin (polymerized) Terpineol Proprietary ingredients

Name	Components
Justrite Thinner and Cleaner	2-Ethoxyethanol
Kaowool	Refractory ceramic fiber (aluminosilicate)
Kester 103 Thinner	n-Propyl alcohol Isopropyl alcohol
Kester 104 Thinner	sec-Butyl alcohol n-Propyl alcohol
Kester 108 Thinner	Isopropyl alcohol
Kester 145/1585 Rosin Flux	Isopropyl alcohol Rosin
Kester 185 Rosin Flux	Isopropyl alcohol Rosin sec-Butyl alcohol
Kester 2211 Organic Flux	Ethyl alcohol Hydrochloric acid
Kester 4422 Thinner	Ethyl alcohol
Kester 5569 Solder-Nu	Water Thiourea Fluoboric acid
Kester 5612 Protecto	Butyl alcohol Xylene Rosin
Kester 5751 Tinning Oil	Polyoxalkylene glycol
Kester 5760/5761 Neutralizer	Water Ammonium phosphate Sodium carbonate
Kester 5765 Scale-Off	Phosphoric acid

Name	Components
Kodagraph Liquid Developer	Water Potassium sulfate Potassium carbonate Diethylene glycol Hydroquinone
Kodak 33 Stop Bath	Acetic acid
Kodak 55 Developer	Hydroquinone
Kodak 66 Developer	Hydroquinone Potassium hydroxide
Kodak Chromium Intensifier, Part A	Potassium chlorochromate
Kodak Chromium Intensifier, Part B	Sodium metabisulfite
Kodak Ektaprint K/L Developer	Iron powder Polymer Carbon black
Kodak Farmer's Reducer, Part A	Potassium ferricyanide
Kodak Farmer's Reducer, Part B	Sodium thiosulfate
Kodak Micro Positive 934	Water Tetramethyl ammonium hydroxide
Kodak Micro Positive Resist ZX-825	Ethyl 3-ethoxypropionate
Kodak MX-936	2-Ethoxyethyl acetate
Kodak Photoresist Developer	Xylene 2-Methoxyethyl acetate
Kodak Rapid Fixer, Part A	Water Ammonium thiosulfate Sodium acetate Boric acid

Name	Components
Kodak Rapid Fixer, Part B	Water Ammonium sulfate Sulfuric acid
Kodak Sepia Toner, Part B	Sodium sulfide
Kovar Bright Dip (412X)	Acetic acid Nitric acid
Kovar Bright Dip (RDX-555)	Sulfuric acid Nitric acid
Krytox 1500/1600 Series Fluorinated Oil	Hexafluoropropylene epoxide
KTI 45 (−)	Xylene
KTI 732/732R/732-A313A (−)	Xylene 2-Methoxyethanol
KTI 747/747R/747-A313A (−)	Xylene 2-Methoxyethanol
KTI 752 (−)	Xylene 2-Methoxyethanol
KTI 802 Negative Developer	Xylene Petroleum distillates
KTI 809 Positive Developer	Sodium phosphate Tribasic alkaline water solution Sodium silicate
KTI 820 (+)	2-Ethoxyethyl acetate
KTI 820J (+)	2-Ethoxyethyl acetate
KTI 934 Positive Developer	Water Tetramethyl ammonium hydroxide
KTI 1300 Thinner	2-Ethoxyethyl acetate n-Butyl acetate Xylene

Name	Components
KTI 1350J (+)	2-Ethoxyethyl acetate Xylene n-Butyl acetate Ethyl benzene
KTI 1370/1375 (+)	2-Ethoxyethyl acetate Xylene Butyl acetate
KTI 1470 (+)	2-Ethoxyethyl acetate Xylene n-Butyl acetate Ethyl acetate
KTI 9000/9000K/9010 (+)	2-Ethoxyethyl acetate Bis(2-methoxyethyl) ether
KTI II (+)	2-Ethoxyethyl acetate Xylene n-Butyl acetate
KTI II Negative Developer	Stoddard solvent
KTI Aluminum Etch I	Phosphoric acid Water Acetic acid Nitric acid
KTI Aluminum Etch II	Phosphoric acid Acetic acid Water Nitric acid
KTI Aluminum Etch III	Phosphoric acid Water Nitric acid
KTI Buffered Oxide Etch 6 : 1	Ammonium fluoride Hydrofluoric acid
KTI Buffered Oxide Etch 50 : 1	Ammonium fluoride Hydrofluoric acid

Name	Components
KTI Chrome Etch	Ceric ammonium nitrate Nitric acid
KTI COP Developer I	Methyl ethyl ketone Ethyl alcohol
KTI COP Rinse I	Isopropyl alcohol Methyl isobutyl ketone
KTI COP Rinse II	Isopropyl alcohol
KTI DE-3 (+)	Water Sodium hydroxide Sodium silicates Sodium phosphates
KTI DE-5/SB-351 (+)	Water Sodium borate Sodium hydroxide
KTI KPR Negative Developer	Xylene 2-Methoxyethyl acetate
KTI Mask Protective Coating	Water Isopropyl alcohol Polyvinyl alcohol resin
KTI/MEAD COP I (−)	Methyl ethyl ketone Ethanol
KTI NMD-25 (+)	Water Ethylene glycol Ethanolamine Isopropyl alcohol
KTI Oxide Etch 5 : 1	Hydrofluoric acid
KTI Oxide Etch 10 : 1	Hydrofluoric acid
KTI Oxide Etch 50 : 1	Hydrofluoric acid

Name	Components
KTI Pad Etch	Acetic acid Ammonium fluoride Water Aluminum acetate
KTI PBR I Cleaning Compound	Ethyl 3-ethoxypropionate Acetone
KTI PBS Developer II	Methyl isoamyl ketone
KTI PBS Rinse	Isopropyl alcohol Methyl isoamyl ketone
KTI Photoresist Standard (−)	Xylene Ethylbenzene
KTI Photoresist Stripper C.S.	Sulfuric acid Chromic acid
KTI PM Acetate	Propylene glycol monomethyl ether acetate
KTI PMMA Rinse	Isopropyl alcohol Methyl isobutyl ketone
KTI PMMA-Standard 496k/950k	Chlorobenzene
KTI Projection Developer	Mineral spirits
KTI Rinse #1, COP	Isopropyl alcohol Methyl isobutyl ketone
Kwik Dri 66	Stoddard solvent
Laminar Developer KB-1B	2-Butoxyethanol
Laminar Developer KB-113	2-Butoxyethanol
Laminations, Transformer	Iron Silicon

Name	Components
882 Laser Gas	Helium Carbon dioxide Nitrogen Carbon monoxide
Laser Gas 920L	Carbon dioxide Carbon monoxide
Lauder Tray Cleaner Concentrate	Potassium dichromate Sulfuric acid
Lea Ronal NP-A and NP-B Solder Stripper	Hydrogen peroxide Salts of hydrofluoric acid
Liquichlor	Sodium hypochlorite
Liquid Alkaline Strip 7463	2-Ethoxyethyl acetate Alkaline organic solvents Hydroxides
Lithium Batteries	Lithium tetrachloride
Locotenge 520	Ethanolamine 2-Butoxyethanol
Locquic Primer N	1,1,1-Trichloroethane
Losolin IV	Naphtha Dodecylbenzene sulfonic acid
Low Dye — Fast Dry Ink	Ethylene glycol Acid dye (suspected carcinogen) Formaldehyde
3M Scotchkote Electrical Coating	Acetone Methyl ethyl ketone Toluene Elastomer Resin Glycerol esters Zinc oxide Antioxidant

Name	Components
MAE Etchants	Nitric acid Hydrofluoric acid Acetic acid
Magic Glass Cleaning and Anti Fogging Fluid	Isopropyl alcohol Surfactant Glycerin Antistatic compound
Magnifloc 509C Flocculant	Formaldehyde Hydrogen chloride
Magnifloc 1561C Flocculant	Stoddard solvent
Magnifloc 1820A Flocculant	Stoddard solvent
Markem 215 Cleaner	Ethyl alcohol
Markem 320 Cleaner	Isopropyl alcohol Ethyl acetate
Markem 452 Washout Solution	Ethyl alcohol
Markem 500 Cleaner	Ethylene glycol monopropyl ether Sodium metasilicate pentahydrate
Markem 501 Cleaner	Ethyl alcohol Water Cresol
Markem 531 Gravure Plate Stripper	Methylene chloride Trichlorotrifluoroethane (Freon)
Markem 535 Cleaner	Ethanolamine
Markem 540 Cleaner	Ethylene glycol phenyl ether
Markem 6993 Black	Ethylene glycol phenyl ether
Markem 6993 Red 6-11C	Ethylene glycol phenyl ether
Markem 7224 Ink Series	Tributyl phosphate
Markem 7254 Ink Series	Triethylene glycol monomethyl ether

Name	Components
Markem 9040 UV Curable Ink	Epoxy resins
Markem 9060 Ink Series	Tri/tetra acrylate esters of pentaerythritol
Markem 9060 White	Pentaerythritol triacrylate Dimethoxy phenylacetophenone Trimethylolpropane triacrylate
Markem 9081 White	Propylene carbonate
Markem Thinner A	Esterified glycol ether
Markem Thinner O	Glycol ether
Markem Thinner XF	2-Ethoxyethyl acetate
MC Cleaner, ZC-675	1,1,1-Trichloroethane
Melamine Mold Cleaning Compound 158/92	Melamine-formaldehyde resin
Merigraph Post — Ext #1	Potassium persulfate
Metal Etch (KTI Aluminum Etch)	Phosphoric acid Acetic acid D.I. water Nitric acid
Metalex Tin Stripper L-2	Sodium hydroxide
Metalex W Special	Sodium hydroxide
Metasil-11	Sodium silicate
Metex Copper Addition Agent S-1	Acetylenic glycol Potassium hydroxide Tellurium compound
Metex S-1699	Glycols
Metex Tin Stripper L-2	Sodium hydroxide
MF-312 Developer	Tetramethyl ammonium hydroxide

Name	Components
Micropolish Alumina	Aluminum oxide
Microposit 111S (+)	2-Ethoxyethyl acetate n-Butyl acetate Xylene Toluene
Microposit 119S (+)	2-Ethoxyethyl acetate Chlorotoluene n-Butyl acetate Xylene Toluene
Microposit 119 Thinner	2-Ethoxyethyl acetate o-Chlorotoluene n-Butyl acetate Xylene Toluene
Microposit 303A/351	Sodium hydroxide
Microposit 351 (Developer)	Sodium hydroxide
Microposit 352/353 (Developer)	Proprietary ingredients Sodium hydroxide
Microposit 354 (Developer)	Sodium hydroxide
Microposit 452	Water Potassium hydroxide
Microposit 452/454/455 (Developer)	Potassium hydroxide
Microposit 1375 (+)	2-Ethoxyethyl acetate n-Butyl acetate Xylene
Microposit 1400-33 (+)	2-Ethoxyethyl acetate n-Butyl acetate Xylene

Name	Components
Microposit 1400S (+)	2-Ethoxyethyl acetate n-Butyl acetate Xylene
Microposit 1450J (+)	2-Ethoxyethyl acetate n-Butyl acetate Xylene
Microposit 1470 (+)	2-Ethoxyethyl acetate Xylene Butyl acetate
Microposit CD-31 (Developer)	Water Alkaline salts, principally phosphates
Microposit Developer Concentrate CD-30	Water Alkaline salts, principally phosphates
Microposit Developer Sal 101	Proprietary ingredients Tetramethyl ammonium hydroxide
Microposit Developer XP-5022	Proprietary ingredients Tetramethyl ammonium hydroxide
Microposit EBR-10	Propylene glycol monomethyl ether acetate
Microposit MF 312/314/319/322 (Developer)	Tetramethyl ammonium hydroxide
Microposit MF 320/321 (Developer)	Proprietary ingredients Tetramethyl ammonium hydroxide
Microposit NPE-210 Solution	Isopropyl alcohol Tetramethyl ammonium hydroxide
Microposit Remover 140	p-Toluene sulfonic acid

Name	Components
Microposit Remover 1112A	Glycol ethers 2-Butoxyethanol Ethanolamine Dipropylene glycol methyl ether Furfuryl alcohol
Microposit Remover 1165	M-Pyrol
Microposit S1813 (+)	Propylene glycol monomethyl ether acetate
Microposit S1817 (+)	Propylene glycol monomethyl ether acetate
Microposit S1818 (+)	Propylene glycol monomethyl ether acetate
Microposit S1822 (+)	Propylene glycol monomethyl ether acetate
Microposit Sal 601-ER7 (+)	2-Ethoxyethyl acetate n-Butyl acetate Xylene
Microposit Surface Coating FSC-L/FSC-M (+)	Propylene glycol monomethyl ether acetate
Microposit Thinner Type P	Propylene glycol monomethyl ether acetate
Microposit XP-2144	Tetramethyl ammonium hydroxide
Microposit XP-5016-24	Tetramethyl ammonium hydroxide
Microposit XP-6009 (+)	2-Ethoxyethyl acetate n-Butyl acetate Xylene
Microposit XP-6012 (+)	2-Ethoxyethyl acetate n-Butyl acetate Xylene

Name	Components
Microposit XP-8807-J1 (+)	Dipropylene glycol methyl ether acetate
Microstrip	Cresol Alkyl aryl sulfonic acid Chlorinated solvents
Microstrip 2001	Cyclic N substituted amide Substituted aliphatic amine
Milros 611 Flux	Alcohol White gum rosin
Mixed Acid Etch (5-2-2)	Nitric acid 5 ⎫ Acetic acid 2 ⎬ ratio Hydrofluoric acid 2 ⎭
Mixed Acid Etch (6-1-1) (Stress Relief Etch)	Nitric acid 6 ⎫ Acetic acid 1 ⎬ ratio Hydrofluoric acid 1 ⎭
MS-111 Epoxy Stripper	Methylene chloride Formic acid Phenol Surfactants
MS-114 Conformal Coating Stripper	Trichlorotrifluoroethane (Freon) Tertiary amine solvent (e.g., nitromethane) Acetone
MS-190 Flux Remover	Methylene chloride Trichlorotrifluoroethane (Freon) Dichlorodifluoromethane (Freon)
MS-190 HD Flux Remover	Methylene chloride Trichlorotrifluoroethane (Freon) Dichlorodifluoromethane (Freon)
MS-460 Conformal Coating	Trichlorofluoromethane (Freon) Dichlorodifluoromethane (Freon)

Name	Components
MS-462 Conformal Coating	Toluene
MS-470 Urethane Coating	Methylene chloride Dichlorodifluoromethane (Freon) Trichlorofluoromethane (Freon) 2-Ethoxyethyl acetate Xylene Toluene-2,4-diisocyanate
MS Cleaner	Ethyl alcohol 2-Butoxyethanol Hydrocarbon propellant
MU9	Methylene chloride Methyl alcohol
Nanostrip	Sulfuric acid Hydrogen peroxide Surfactant
Nayacol C-38/C-58 Colloidal Silica Solution	Silicon dioxide
Neutronex 309 Grain Refiner	Thallium
NF Solder Stripper 3114B	Nitric acid Ammonia
Niklad 794-A	Water Nickel
Niklad 794-B/794-HZ	Water Sodium hypophosphite
2161 Organic Flux, Kester	n-Propyl alcohol
Oakite 33	Phosphoric acid 2-Butoxyethanol Ethoxylated nonyl phenol
OFPR-800 (+)	2-Methoxyethyl acetate

Name	Components
OFPR-800 AR-15 (+)	2-Ethoxyethyl acetate
OFPR Developer DE-3	Inorganic silicates
OFPR Developer NMD-3	Alkyl ammonium hydroxide
Omega Meter Solution	Isopropyl alcohol
OMR-83F (−)	Xylene
Opti Skan Scan Cleaner	Isopropyl alcohol Polydimethylsiloxane
Opto Kleansor G-091-06	2-Methoxyethanol
Organo Flux 3355-11	Isopropyl alcohol Hexylene glycol
P12571D	M-Pyrol Aromatic 100 (petroleum distillate) Polyamic acid of pyromellitic dianhydride/4,4-oxydianiline (polymer) Polyamic acid of benzophenone Methanol Proprietary ingredients Tetracarboxylic dianhydride/4,4-oxydianiline/m-phenylenediamine (polymer)
P12572	M-Pyrol Aromatic 100 (petroleum distillate) Polyamic acid of pyromellitic dianhydride/4,4-oxydianiline (polymer) Polyamic acid of Benzophenone Methanol Proprietary ingredients Mineral spirits

Name	Components
Pad Etch (Vapox Etch)	Acetic acid D.I. water Ammonium fluoride Aluminum acetate
Passivation Solution	Nitric acid
Patclin 948 Solder Stripper	Hydrogen peroxide Salts of hydrofluoric acid
Patclin 958	Phosphoric acid Nitric acid
PBS Developer	Methyl isoamyl ketone Isopropyl alcohol
PBS Rinse	Isopropyl alcohol Methyl isoamyl ketone
PC-96 Solvent Soluble Resist	1,1,1-Trichloroethane Isopropyl alcohol Resin
PD Activator	Ethylene glycol Ethanolamine 2-(2-Aminoethoxy) ethanol Hexylene glycol Benzyl alcohol Oleic acid
PD-86 Developer	Hydroquinone
PDE-100	Carbon tetrafluoride Oxygen
Perma Dry Ink 555 Red	Diacetone alcohol Resins Lead compounds 2-Butoxyethanol Color

Name	Components
Perma Dry Ink 555 White	Diacetone alcohol Pigments Resins
PFC	Acetic acid
PF Etchant	Ferric chloride
PF-95 Fixer	Ammonium thiosulfates
P10 Gas	Argon Methane
Phosphorous Planar Diffusion Sources	Silicon phosphorus
Phosver 3 Power Pillows	Potassium pyrosulfate
Photoposit 7411 Defoamer	Stoddard solvent Trialkyl phosphates
PI 2563	M-Pyrol Xylene 2-Ethoxyethanol Methyl alcohol Ethyl alcohol
PIQ 13	M-Pyrol Dimethyl acetamide
PIQ Coupler	Toluene
PIQ Coupler 3	M-Pyrol
PIQ Etchant	Hydrazine hydrate
Piranha Etch	Sulfuric acid Hydrogen peroxide
PIX-1400X-1	M-Pyrol
PIX-3500	M-Pyrol

Name	Components
PM-940 Etch	Chromium, hexavalent (from chromic acid)
PM-950 Neutralizer	Amines (including diethylene triamine)
Poly-6	M-Pyrol
Poly Etch 95%	Nitric acid Acetic acid Hydrofluoric acid
Poly I Gas	Chlorine Oxygen Nitrogen Carbon dioxide
Poly II Gas	Carbon monoxide Carbon tetrafluoride Carbon dioxide Nitrogen Sulfur hexafluoride
Polyimide Coating PI 2545	M-Pyrol Aromatic hydrocarbons Resin
Polyimide Coating PI 2561	M-Pyrol Resin
Posistrip 830 Stripper	2-(2-Aminoethoxy) amine M-Pyrol
PPD 5932 Developer	Ammonium hydroxide
PR-21 Resist	2-Ethoxyethyl acetate n-Butyrolacetone
PR-55 Resist	2-Ethoxyethyl acetate n-Butyrolacetone

Name	Components
Pre-Metal Etch (Poly Silicon Etch)	Ammonium fluoride Water Ammonium phosphate Ammonium hydroxide
Prepared Bath 2137	Fluoboric acid Tin
Preposit Etch 746/748	Sulfuric acid
Pre-Sputter Clean	Ammonium fluoride Ammonium phosphate
Pre-Sputter Etch	Ammonium fluoride Water Ammonium phosphate
Primer 910S	Isopropyl alcohol
Probe RK	2-Butoxyethanol Sodium metasilicate
Processor Fixer Concentrate	Acetic acid
Pro 330 Clear Thin Spread	Petroleum oil Mineral spirits Ethylene glycol Ammonia
Protecto 5612	Butyl alcohol Xylene Rosin
PRS-2000 Cleaner	M-Pyrol Sulfolane 2-(2-Ethoxyethoxy) ethanol Tetraethylene glycol Monoisopropanol amine
PRS-3000 Cleaner	M-Pyrol Sulfolane Monoisopropanol amine

Name	Components
Pyralin PI 2563	M-Pyrol Polyimide resin 2-Ethoxyethanol Xylene Methyl alcohol Ethyl alcohol
Pyralin PI 2570	M-Pyrol Aromatic hydrocarbons Polyamide polymeric resin Toluene Proprietary ingredient(s)
Q-Level Brightener	Potassium cyanide Potassium hydroxide Lead acetate
QX 3301 Polyimide Developer	Aliphatic ester Xylene
QZ 3289/3290 Adhesion Concentrate	Ethyl alcohol
QZ 3312 Rinse Two	Xylene
197 Rosin Flux	Isopropyl alcohol Rosin
1500 Replenishing Salts	Potassium cyanide
Rapid Film Fix	Acetic acid
RCA Clean (Step 1)	Water Hydrogen peroxide Ammonium hydroxide
RCA Clean (Step 2)	Water Hydrogen peroxide Hydrochloric acid
RDH Lime Solvent	Phosphoric acid Acetic acid

Name	Components
RDX-221 Alumetch LK	Hydrofluoric acid
Red Epoxy Ink (RINK)	Diacetone alcohol Red dye
Red Fuming Nitric Acid (RFNA)	Nitric acid Nitrogen dioxide
Red Glpt Varnish 10-9002/9003/9008/9009	Xylene Iron oxide Resin
Reducer DTR602	Naphtha Toluene Acetone Alkyl benzenes 2-Butoxyethyl acetate
Repromatic 140 Short Stop	Acetic acid
Riston II Antifoam 200	Octyl alcohol
Riston II D200 Developer	2-Butoxyethanol Sodium tetraborate decahydrate
Riston II Stripper Antifoam	Isopropyl alcohol
Riston II Stripper S-1000X Concentrate	2-(2-Butoxyethoxy) ethanol Ethanolamine
Ritchpak-40, A-Side	Methylene bisphenyl isocyanate monomer Methylene bisphenyl isocyanate polymer Trichlorofluoromethane
Ritchpak-40, B-Side	Trichlorofluoromethane Tertiary amine catalyst

Name	Components
Ritchpak-42, A-Side	Methylene bisphenyl isocyanate monomer Methylene bisphenyl isocyanate polymer Trichlorofluoromethane
Ritchpak-50, A-Side	Methylene bisphenyl isocyanate monomer Methylene bisphenyl isocyanate polymer Trichlorofluoromethane
Ritchpak-70, A-Side	Methylene bisphenyl isocyanate monomer Methylene bisphenyl isocyanate polymer Trichlorofluoromethane
RN-10 E-Beam Negative Resist Rinse	Isopropyl alcohol Methyl isobutyl ketone
RN-11 Developer	Isopropyl alcohol
RN-11 E-Beam Negative Resist Rinse	Isopropyl alcohol
Rocheltex	Potassium hydroxide
Rokleen 10-12	Highly alkaline
Rosin Flux	Isopropyl Alcohol Rosin
Rosin Flux Kester #135	Isopropyl alcohol Rosin
Rosin Flux Kester #1544 Mil	Rosin Isopropyl alcohol
Rostrip M-12	Sodium cyanide Sodium hydroxide

Name	Components
RP-10 E-Beam Positive Resist Rinse	Isopropyl alcohol Methyl isoamyl ketone
RS-4 Stripper, Positive	2-Butoxyethanol Sodium hydroxide
R10 Stripper	Diethylene glycol monobutyl ether Ethanolamine
RT-2 Stripping Solution	Sulfuric acid Nitric acid Chromic acid
RTV 112 Silicone Sealant	Methyltriacetoxysilane Proprietary ingredient
RTV 159 Silicone Sealant	Methyltriacetoxysilane Proprietary ingredient
RTV 851A Black Polysiloxane Foam	Polysiloxane
RTV 851B Silicone Foaming Agent	Polysiloxane
RTV 7403 Silicone Sealant	Silicone
RTV Sealant 732	Acetoxysilane
Rustlick B-5J	Ethanolamine Sodium chromate Diethanolamine
911 Stripper	Formic acid Chromic acid
S-1634	Sodium hydroxide
Scan Kleen	Isopropyl alcohol Ammonium hydroxide
Scrubber-Vapox	D.I. water Ammonium hydroxide

Name	Components
SE30	Polydimethylsiloxane
SE33	Polydimethylsiloxane
Selectilux (−)	Xylene
Selectilux N 35 (−)	Xylene
Selectilux N 45-322 (−)	Xylene
Selectilux N 65 (−)	Xylene
Selectilux P 15 (+)	2-Ethoxyethyl acetate
Selectiplast N2 Negative Developer	C8–C11 aliphatic hydrocarbons (e.g., gasoline, kerosene) C3–C4 alkyl benzenes Xylene
Selectiplast P2 Negative Developer	Sodium silicates Phosphates
Sel-Rex Circuitprep SC Replenisher/Make Up	Ethylenediamine Ethanolamine Cyanide
Sel-Rex Cu Bath M Lo	Sulfuric acid
Sel-Rex Neutronex 309 Brightener	Oxalic acid Thallium
Sel-Rex XR-170A Pretreatment	Ethylenediamine Ethanolamine
Silicon Etch Solution	Nitric acid Acetic acid Hydrofluoric acid
Silver De-Plate Solution	Water Sodium cyanide
Silver Glo 3K	Potassium cyanide Thiouracil

Name	Components
Silver Glo 3KBP/33BP	Water Organic acids Antimony Cyanide
Silver Glo 33	Cyanide salt
Silver Glo BP	Water Organic acids Antimony
Silver Glo TY	Water Surfactant
Silver Jet 300 SD Brightener	Cyanide compound
Silver Jet 300 SD Silver Salts	Potassium silver cyanide
Silver Jet Brightener 100/220	Cyanide salt
Silver Make-Up Solution	Potassium silver cyanide
Silver Plating Solution	Water Potassium cyanide Silver cyanide Potassium carbonate
Silver Replenisher Brightener HCD	Cyanide compound
Silver Salts HDC	Potassium silver cyanide
Simple Green (Sunshine Makers)	2-Butoxyethanol
Sisulfasol	Chlorosulfonate Methylene chloride
Slago (AM Solder Flux)	Ammonium chloride Stannic chloride Boron trioxide
Slotelet Additive K	Methyl alcohol

Name	Components
Slotelet G10 Tin/Lead Plating Solution	Alkyl sulfonic acid
Slotelet K Tin/Lead Plating Solution	Alkyl sulfonic acid
Snoop (Nupro)	Water Halides Surfactant
Soda Lime	Calcium oxide Sodium hydroxide (or potassium hydroxide)
Solder	Tin Lead
Solder Flux	Isopropyl alcohol Ethyl alcohol Methyl alcohol Terpene resins
Solder Flux, 1429 Organic	Water Glutamic acid hydrochloride Urea
Solder Flux, 2163 Organic	Isopropyl alcohol Proprietary organic acid Proprietary salt Proprietary surfactant Water
Solder Flux Thinner	Isopropyl alcohol Ethyl alcohol Methyl alcohol Terpene resins
Solder Paste Flux	Lead/tin solder Glycol ether
Solderon Acid	Organic acid mixture

Name	Components
Solderon Lead Concentrate	Lead salt organic acid
Solderon SG Additive	Methyl alcohol
Solderon SG Make Up	Carboxylic acid salts
Solderon Tin Concentrate	Tin salt organic acid
Solder Strip NP-A	Nitric acid Organic sulfonic acid
Solder Strip 8T	Fluorides
Solder Stripper 948	Hydrogen peroxide salts of hydrofluoric acid
Solvent 111	1,1,1-Trichloroethane Dioxane
Space Tuner Cleaner	Trichlorofluoromethane (Freon MS) Lubricant
Spinrite Arsenic	Isopropyl alcohol Silica Arsenic trioxide
Spruddle/Puddle	Water Tetramethyl ammonium hydroxide Tergitol
Stafilm 1020 Die Bonding Adhesive	Silver Phenyl glycidyl ether
Starter 2000/2137	Fluoboric acid
Staticide (Analyt Chem Lab)	Quaternary ammonium salts
Stay-Sily White Flux	Boric acid
Strip L	Methylene chloride Phenol Acid activators Wetting agents

Name	Components
Stripper #3 (Allied Chem)	Sulfuric acid Chromic acid
Stripper #5 (Allied Chem)	Sulfuric acid Chromic acid
Striptex Mu-7 Resist Remover	Methylene chloride Methyl alcohol
Sumikon EME 6300/6300H/6300HC/8150 (Epoxy Molding Compounds)	Silica (fused) Resins, phenolic and epoxy Antimony trioxide Carbon black
Sumikon EME 9100X (Epoxy Cresol Novolac Resin System)	Silica Carbon black Catalyst
Sumikon EME 9300H	Silica (fused) Resins, phenolic and epoxy Antimony trioxide Carbon black
Sunny Sol 100/150	Sodium hypochlorite Sodium hydroxide
Super Imped	Sodium hypochlorite
Super Q Etch (Image Tech)	Ammonium fluoride Phosphoric acid
Super Solder Strip 1805 (J & S Labs)	Ammonia Hydrogen peroxide
Super Strip 100	Aromatic nitro sulfonic acid Salt (alkaline salt)
Surdip Striper RDX-522	Organic acids Tetrachloroethylene
Surfynol 104PA Surfactant	Isopropyl alcohol

Name	Components
System MFI-1 Metal-Ion-Free Immersion Developer	Tetramethyl ammonium hydroxide
Technilead NF Concentrate	Lead salt Organic sulfonic acid
Technistrip Au	Potassium cyanide Lead oxide
Teflon Dry Lube 708	Methylene chloride 1,1,1-Trichloroethane Propane
Tergitol Nonionic Surfactants	Polyethylene glycol ether
Thinner E	2-Ethoxyethyl acetate
Tin Glo-Culmo Starter/Brightener	Methyl alcohol
Top Dip	Phosphoric acid
TPD Thinner	1,1,1-Trichloroethane Toluene Methylene chloride
TR-240A/240IMA/250A/257A	Methylene bisphenyl isocyanate High molecular weight polymers Trichlorofluoromethane (Freon)
Triton-X-100	Alkyl aryl polyether alcohol (surfactant)
Trycol 6720	Alkoxylated alcohol
Tryfac 5568/5569-A	Phosphated aryl ethoxylate
TSMR 8800 (+)	2-Ethoxyethyl acetate n-Butyl acetate Xylene

Name	Components
TSMR-8800 BE	Ethyl lactate Novolak resin Butyl acetate
Twinkle Stainless Steel Cleaner	1,1,1-Trichloroethane Propane Butane
UCAR PM Acetate	Propylene glycol monomethyl ether acetate
Ultramac 55	2-Ethoxyethanol Resin
Ultramac D89	Sodium hydroxide
Ultramac EPA 1024 MB-627AN (Resin)	Ethyl lactate
Ultramac MF29	Alkanolamine Tetramethylammonium hydroxide
Ultramac MF62	Amine mixtures Alkanolamine Tetramethylammonium hydroxide
Ultramac MF62A/MF81A	Alkanolamine Tetramethylammonium hydroxide
Ultramac MF81A-47% W/W	Alkanolamine Tetramethylammonium hydroxide
Ultramac PR68 (Resin)	2-Ethoxyethyl acetate Resin Solvent blend
Ultramac PR-1024 MB-628 (Resin)	2-Ethoxyethyl acetate n-Butyl acetate Xylene

Name	Components
Ultramac S40	Dichlorobenzene Alkyl aryl sulfonic acid Phenol
Ultramac S41	Dipropylene glycol monomethyl ether
Ultramac S43/S46	Surfactant
Ultramac Solvent EPA	Xylene n-Butyl acetate
Uresolve Plus and SG	2-Methoxyethanol Wetting agents Color Oxygenated solvents Strong base
Uvex Primer 910S	Isopropyl alcohol
V-183	Toluene Resin
Vandalex 20/24	Toluene Isopropyl alcohol Methyl ethyl ketone 1,1,1-Trichloroethane Ethylene glycol Carbon dioxide
VWR Glass Cleaner	Isopropyl alcohol Propyl alcohol Surfactant
Waxivation Compound	Acetone Glycol phthalate 2-Ethoxyethyl acetate
Waycoat 28 (−)	Xylene Ethyl benzene

Name	Components
Waycoat 43 (−)	Xylene Ethyl benzene
Waycoat 59 (−)	Xylene Ethyl benzene
Waycoat 204 (+)	2-Ethoxyethyl acetate n-Butyl acetate Xylene Ethyl benzene
Waycoat Developer (−)	Xylene Ethyl benzene
Waycoat HNR 80 (−)	Xylene Ethyl benzene
Waycoat HNR 120 (−)	Xylene Ethyl benzene
Waycoat HPR 205/207 (+)	2-Ethoxyethyl acetate n-Butyl acetate Xylene Ethyl benzene
Waycoat HPR 504/505 (+)	Ethyl lactate
Waycoat HPRD-402/407/428/429 (+)	Tetramethyl ammonium hydroxide
Waycoat HR100 (−)	Xylene
Waycoat HR200 (−)	Xylene Ethyl benzene
Waycoat Negative Resist Developer	Isoparaffinic hydrocarbons
Waycoat PF Developer	Xylene Stoddard solvent

Name	Components
Waycoat RX 507 (+)	2-Ethoxyethyl acetate n-Butyl acetate Xylene Ethyl benzene
Waycoat SC100 (−)	Xylene Ethyl benzene
Waycoat SC100 CP (+)	Xylene Ethyl benzene
Waycoat SC180 (−)	Xylene Ethyl benzene
WCD 10BL (Corrosion Inhibitor)	Hydrogen peroxide
Weld-On 3 for Acrylic	Methylene chloride Trichloroethylene Methyl methacrylate (monomer)
Weld-On 711	Tetrahydrofuran Dimethylforamide Resin (with fillers and pigments)
Weld-On P-70 Primer	Tetrahydrofuran Methyl ethyl ketone Cyclohexanone Dimethylformamide
Weld-On P-72 Primer	Tetrahydrofuran
Wet K-Etch	Phosphoric acid D.I. water Acetic acid Nitric acid Ammonium fluoride 13 : 2 BOE
Whirlwind Glass Cleaner	Isopropyl alcohol 2-Butoxyethanol

Name	Components
Wright Etch	Hydrofluoric acid Acetic acid Copper nitrate Chromic acid Nitric acid
WRS 200S Solution	Water Isopropyl alcohol Organic ammonium salt
WX-104 Positive Photoresist Stripper	Aliphatic amine Cyclic amide
WX-427 Developer (+)	Trisodium phosphate Sodium metasilicate
Xactseal	Xylene
Xanthochrome (+)	2-Ethoxyethyl acetate n-Butyl acetate Xylene Ethyl benzene
Xerox 1020/1035/2830 Developer	Iron powder Styrene, acrylate copolymer Acrylic resin Polyolefin Silica, amorphous
Xerox 1020/1035/2830 Toner	Styrene, acrylate copolymer Acrylic resin Carbon black Polyolefin Silica, amorphous

Name	Components
Xerox 1040/1045/1048 Developer	Iron powder Styrene, acrylate copolymer Carbon black Polymethylmethacrylate Silica, amorphous
Xerox 1040/1045/1048 Toner	Styrene, acrylate copolymer Carbon black Silica, amorphous
Xerox 1075/1090/4050 Developer	Steel powder Styrene, acrylate copolymer Carbon black Quaternary ammonium salts
Xerox 1075/1090/4050 Toner	Styrene, acrylate copolymer Carbon black Quaternary ammonium salts
Xerox CE-Cyan/ECP-Cyan Premix	Isoparaffinic petroleum solvent Methacrylate copolymer Cyan dye
Xerox CE-Magenta/ECP-Magenta Premix	Isoparaffinic petroleum solvent Methacrylate copolymer Magenta dye
Xerox CE-Process/ECP-Process Black Premix	Isoparaffinic petroleum solvent Methacrylate copolymer Carbon black
Xerox CE-Yellow/ECP-Yellow Premix	Isoparaffinic petroleum solvent Methacrylate copolymer Yellow dye
Xerox Cleaner, Formula A	Water Isopropyl alcohol Butoxypropanol Surfactant Sodium xylenesulfonate Tetrapotassium pyrophosphate

Name	Components
Xerox ECP-Cyan Toner Concentrate	Isoparaffinic petroleum solvent Methacrylate copolymer Blue dye
Xerox ECP-Magenta Toner Concentrate	Isoparaffinic petroleum solvent Methacrylate copolymer Magenta dye
Xerox ECP-Process Black Toner Concentrate	Isoparaffinic petroleum solvent Methacrylate copolymer Carbon black
Xerox ECP-Yellow Toner Concentrate	Isoparaffinic petroleum solvent Methacrylate copolymer Yellow dye
Xerox Film Remover, Tip Wipes	Isopropyl alcohol
XIR-3000-T (Resin)	n-Butyl acetate Xylene
ZC-211 Lubricant	1,1,1-Trichloroethane Organosilicone, nonionic
Zep Aerosolve	1,1,1-Trichloroethane
Zepelec	Trichlorotrifluoroethane
Zep HD	Sodium borate
Zep Once Over Wall Cleaner	2-Butoxyethanol

11.0 Chemical Hazards and Precautions

11.1 Format

Information in this section is presented in the following format:

Chemical Name [CAS number]

Chemical Formula **HAZARD CLASS**
Synonyms

HAZARDS

Physical State   Data on the chemical (solid, liquid, or gas) are provided here.

Dust/Mist/Gas/Vapor   Data on the airborne chemical are provided here.

TLV/PEL	Threshold Limit Value/Permissible Exposure Limit — whichever is less (1995–1996 TLVs used)
STEL	Short-Term Exposure Limit
IDLH	Immediately Dangerous to Life or Health concentration
VP	Vapor Pressure in millimeters of mercury at 20°C
FP	Flash Point
AT	Autoignition Temperature
ER	Explosive Range **NFPA Diamond**

Important Hazard   Hazards of particular concern are described here. Emphasis is placed on hazards specific to semiconductor operations. Not all hazards are listed.

Pictograms   Symbols representing the hazards of the chemical are shown here (see Figure 4).

PRECAUTIONS

Precautions to be taken in the handling of the chemical are discussed here. Also, pictograms representing the precautions are shown (see Figure 4).

Hazard Pictograms

| CORROSIVE | POISON | FLAMMABLE | COMBUSTIBLE | OXIDIZER | CANCER CAUSING |

| GAS HAZARD | SPECIAL HAZARD | AIR REACTIVE |

Precaution Pictograms

| WATER REACTIVE | NO IGNITION SOURCE | FACESHIELD AND GLASSES | APRON AND ARMGUARDS | SOLVENT GLOVES | ACID GLOVES |

| SAFETY GLASSES | AIR EXHAUSTED |

Figure 4. Pictograms

Chemical Name The common name and major synonyms for a chemical are included in this section. Chemical formulas for compounds with simple structures are also included.

Hazard Class Refers to the major hazards of the chemical and is based primarily on NFPA 704. In instances in which a material has more than one major hazard, the primary hazard is listed first. Example: phosphine is both a poisonous gas and a flammable gas. Since the poisonous properties of phosphine are considered the primary hazard, it is listed first as a poison and second as a flammable.

Hazards In most cases, the physical characteristics of compounds listed in this book are those of the pure compound at room temperature. However, in certain instances more than one physical form is listed — e.g., hydrochloric acid (liquid and compressed gas).

When available, the odor threshold of the compound has been included. Because of biological variability, the odor threshold may differ from one person to another. Therefore, a range of odor thresholds is often given. For some chemicals the odor provides good warning properties (e.g., acetic acid, ammonia, hydrochloric acid); for some chemicals it does not. For example:

- Chemical odors can be masked or hidden by other odors in the area.

- Extended exposure to certain chemical odors can decrease an individual's ability to detect these odors (i.e., olfactory fatigue).

- The chemical has no odor or an extremely high odor threshold (e.g., arsenic, carbon monoxide, silver).

- The chemical's odor threshold is higher than the TLV (e.g., arsine, diborane, chloroform).

Table 2 lists common sources of odors. Section 11.2 gives more detailed information on the odor thresholds of specific chemicals.

Table 2. Potential Sources of Odors

Odor Source	Description	IDLH Level (ppm)	TLV Level (ppm)	Odor Level (ppm)
Phosphorus oxychloride	⎫	—	0.1	—
Chlorine	⎪	25	0.5	0.1
Hydrochloric acid	⎪	100	5	1
Aqua regia	⎬ Sharp acidic or bleachy	100	—	1
Nitric acid	⎪	100	2	1
Dichlorosilane	⎪	—	—	—
Silicon tetrachloride	⎪	—	—	—
Boron trichloride	⎭	—	—	—
Acetic acid	Vinegar	1000	10	0.07
Ammonia	Pungent	500	25	5
Ozone	Electric spark	10	0.05	0.02
Phenol	Medicinal	100	5	0.06
Acetone	Nail polish	20,000	750	60
n-Butyl acetate	Sweet	10,000	20	0.3
Xylene	Sweet	10,000	100	3
Isopropyl alcohol	Sweet	20,000	400	20
Methyl alcohol	Sweet	25,000	200	160
Natural gas	Mercaptan (skunk)	20,000	N.A.	1
Gasoline		5000	300	—
Fiberglass resin	Natural gas/aromatic	5000	50	0.05
Auto exhaust	Various	1500	—	<1
Diesel exhaust	Various	50	—	<1
Cafeteria kitchen	Food odors	N.A.	N.A.	—
Paints/varnishes	Sweet solvent	—	—	—
Roofing tar		—	—	—

N.A. — Not available.

11.2 Generic Chemicals

Acetic Acid [64–19–7]

CH₃COOH
Glacial Acetic Acid

**CORROSIVE (ACID)/
COMBUSTIBLE**

HAZARDS

Liquid Can cause chemical burns to skin and eyes. Solutions in water (4–10% acetic acid) are called vinegar.

Vapor Vinegar odor detectable at 0.2 to 1 ppm. Extreme eye and nose irritation occurs at concentrations in excess of 25 ppm. High concentrations can cause coughing, chest pains, nausea, and vomiting. Combustible.

TLV/PEL 10 ppm	**VP** 11 mm	**ER** 5.4 – 16%
STEL 15 ppm	**FP** 109°F	
IDLH 50 ppm	**AT** 869°F	

CORROSIVE   COMBUSTIBLE

PRECAUTIONS

Avoid skin and eye contact. Use solvent gloves, faceshield and safety glasses, apron and armguards. Use only within the confines of an air-exhausted hood.

Keep away from open flame and high temperatures.

FACESHIELD AND GLASSES   SOLVENT GLOVES   APRON AND ARMGUARDS   AIR EXHAUSTED   NO IGNITION SOURCE

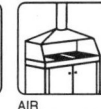

Acetone

[67–64–1]

$(CH_3)_2CO$
Dimethyl Ketone

FLAMMABLE

HAZARDS

Liquid Repeated contact can cause skin and eye irritation.

Vapor Odor detectable at 3.6 to 650 ppm. At approximately 1000 ppm, irritation of the nose and throat can occur. High concentrations (12,000 ppm) can cause headache, drowsiness, weakness, nausea, a feeling of drunkenness, and vomiting. Flammable.

TLV/PEL	750 ppm	**FP**	1.4°F
STEL	1000 ppm	**AT**	869°F
IDLH	2500 ppm	**ER**	2.6 – 13%
VP	266 mm		

FLAMMABLE

PRECAUTIONS

Avoid skin and eye contact. Use neoprene or butyl solvent gloves, faceshield and safety glasses, apron and armguards. Use only within the confines of an air-exhausted hood.

Keep away from ignition sources such as flame, spark, and heat.

FACESHIELD AND GLASSES | SOLVENT GLOVES | APRON AND ARMGUARDS | AIR EXHAUSTED | NO IGNITION SOURCE

Acetonitrile

[75-05-8]

CH₃CN
Methyl Cyanide

FLAMMABLE

HAZARDS

Liquid Can be readily absorbed through the skin. Repeated contact may cause skin and eye irritation.

Vapor Odor detectable at 40 to 500 ppm. Can cause headache, dizziness, drowsiness, and nausea. Higher concentrations can cause vomiting, chest pain, weakness, and stupor. Extremely high concentrations can cause convulsions, coma, and death. Repeated exposure to high concentrations may cause liver, kidney, blood, and testicular damage. Flammable.

TLV/PEL	40 ppm	**VP**	73 mm	**ER**	4.4 – 16%
STEL	60 ppm	**FP**	42°F		
IDLH	500 ppm	**AT**	975°F		

Important Hazard When heated to decomposition, emits highly toxic fumes of cyanide, and oxidizes readily in air to form reaction products that may explode spontaneously.

FLAMMABLE SPECIAL HAZARD

PRECAUTIONS

Avoid skin and eye contact. Use solvent gloves, faceshield and safety glasses, apron and armguards. Use only within the confines of an air-exhausted hood.

Keep away from ignition sources such as flame, spark, and heat.

FACESHIELD AND GLASSES   SOLVENT GLOVES   APRON AND ARMGUARDS   AIR EXHAUSTED   NO IGNITION SOURCE

Acetylene

[74–86–2]

C_2H_2

FLAMMABLE GAS

HAZARDS

Gas Odor detectable at 220 to 2600 ppm. Acts as a simple asphyxiant by displacing oxygen in air. Crude acetylene and some commercial grades contain traces of toxic impurities that have been reported as the cause of some injuries or death following acetylene exposure.

AT 581°F

ER 2.5 – 82%

Important Hazard Extremely flammable.

FLAMMABLE   SPECIAL HAZARD   4/0/3

PRECAUTIONS

Use in a well-ventilated area. Store cylinders upright in a cool dry place — under 100°F recommended — away from sunlight.

Except when welding, keep away from ignition sources such as flame, spark, and heat.

SAFETY GLASSES   NO IGNITION SOURCE

Alkyl Benzenes
Aromatic Hydrocarbons **FLAMMABLE**

HAZARDS

Liquid Can be readily absorbed through the skin. Repeated contact can cause skin and eye irritation.

Vapor Odor detectable at low concentrations. Can be irritating to eyes and lungs. High concentrations may cause headache, dizziness, nausea, loss of appetite, fatigue, and liver and kidney damage. Flammable or combustible depending on specific alkyl benzene.

TLV/PEL 25 ppm (as trimethylbenzene)

FLAMMABLE/COMBUSTIBLE

(as Trimethylbenzene)

PRECAUTIONS

Particular care should be taken to avoid skin and eye contact. Use solvent gloves, faceshield and safety glasses, apron and armguards. Use only within the confines of an air-exhausted hood.

Keep away from ignition sources such as flame, spark, and heat.

FACESHIELD AND GLASSES | SOLVENT GLOVES | APRON AND ARMGUARDS | AIR EXHAUSTED | NO IGNITION SOURCE

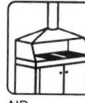

Aluminum, Metal

[7429–90–5]

Al

UNCLASSIFIED

HAZARDS

Dust/Mist Nuisance dust. High concentrations can cause irritation of skin, eyes, nose, and lungs.

TLV/PEL 10 mg/m^3

Important Hazard Powder and flake aluminum are flammable and can form explosive mixtures in air especially when treated to reduce surface oxidation. If fire occurs, smother with inert material or use approved Class D extinguishers; DO NOT use carbon dioxide or halogenated extinguishers; DO NOT use water.

SPECIAL HAZARD

(Aluminum Powder)

PRECAUTIONS

Avoid skin and eye contact. Use acid or solvent gloves and safety glasses. Use in a well-ventilated area.

ACID GLOVES   SAFETY GLASSES

Aluminum Acetate

[142–03–0]

$Al(C_2H_3O_2)_3$

UNCLASSIFIED

HAZARDS

Solid/Liquid Can be irritating to skin and eyes. Repeated or prolonged contact can cause allergic reaction.

Dust/Mist Repeated or prolonged exposure can cause allergic reaction. Can be mildly irritating to eyes, nose, and throat.

TLV/PEL $2mg/m^3$ as Al

PRECAUTIONS

Avoid skin and eye contact. Use solvent gloves and safety glasses.
Use in a well-ventilated area.

SOLVENT GLOVES

SAFETY GLASSES

Aluminum Chloride

[7446–70–0]

AlCl$_3$

CORROSIVE (ACID)

HAZARDS

Dust Can cause burns to skin and eyes. Repeated or prolonged exposure can cause skin rash. Toxicity similar to hydrogen chloride (see Hydrogen Chloride). Reacts violently with water and other solvents.

TLV/PEL 2 mg/m^3 (soluble salts, as aluminum)

CORROSIVE

PRECAUTIONS

Aluminum chloride is present in plasma aluminum etchers and their cold traps. Etcher residues should be kept within the confines of an air-exhausted hood or used with local exhaust ventilation.

Water and aluminum chloride should be mixed slowly.

When working with residues from plasma aluminum etchers, use solvent gloves, faceshield and safety glasses, apron and armguards.

FACESHIELD AND GLASSES | SOLVENT GLOVES | APRON AND ARMGUARDS | AIR EXHAUSTED | WATER REACTIVE

Aluminum Fluoride

[7784-18-1]

AlF$_3$
Aluminum Trifluoride

CORROSIVE (ACID)

HAZARDS

Dust/Mist Can cause burns to skin and eyes. Toxicity similar to hydrogen fluoride (see Hydrogen Fluoride).

TLV/PEL 2 mg/m^3 (soluble salts, as aluminum)
2.5 mg/m^3 (as F)

CORROSIVE

(Manufacturer)

PRECAUTIONS

Avoid skin and eye contact. Use acid gloves, faceshield and safety glasses, apron and armguards. Use only within the confines of an air-exhausted hood.

FACESHIELD AND GLASSES  | SOLVENT GLOVES  | APRON AND ARMGUARDS 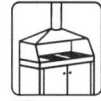 | AIR EXHAUSTED

Aluminum Oxide

[1344–28–1]

Al_2O_3
Alumina

UNCLASSIFIED

HAZARDS

Dust/Mist Nuisance dust. High concentrations can cause irritation of skin, eyes, nose, and lungs. Coughing and shortness of breath can also occur.

TLV/PEL 10 mg/m^3

(Manufacturer)

PRECAUTIONS

Avoid skin and eye contact. Use acid or solvent gloves and safety glasses. Use in a well-ventilated area.

ACID GLOVES   SAFETY GLASSES

Ammonia [7664–41–7]

NH$_3$
Anhydrous Ammonia

CORROSIVE(BASE)

HAZARDS

Liquid In contact with the eyes and skin, can cause serious eye injury and chemical burns. Can be a liquid when under pressure.

Gas/Vapor Pungent, irritating odor detectable at 0.04 to 50 ppm. Irritating to eyes, nose, throat, lungs, and skin. Irritation and discomfort can occur at 20 to 25 ppm. 2500 to 6500 ppm can cause severe eye irritation, difficulty in breathing, chest pain, and fluid in lungs. Combustible gas under certain circumstances.

TLV/PEL	25 ppm
STEL	35 ppm
IDLH	300 ppm
AT	1204°F
ER	16 – 25%

CORROSIVE

PRECAUTIONS

Avoid skin and eye contact. Use acid gloves, faceshield and safety glasses, apron and armguards. Use only in the confines of an air-exhausted hood. Keep away from open flame and high temperatures.

FACESHIELD AND GLASSES | ACID GLOVES | APRON AND ARMGUARDS | AIR EXHAUSTED | NO IGNITION SOURCE

Ammonium Bifluoride

[1341–49–7]

NH_5F_2

CORROSIVE (ACID)

HAZARDS

Solid Can cause chemical burns to skin and eyes.

Dust Can be irritating to eyes, nose, and throat.

TLV/PEL 2.5 mg/m^3 (as fluoride)

CORROSIVE

(Manufacturer)

PRECAUTIONS

Avoid skin and eye contact. Use acid gloves, faceshield and safety glasses, apron and armguards.

Ammonium bifluoride forms when hydrogen fluoride (HF) is mixed with ammonium fluoride (NH$_4$F), as in buffered oxide etch (BOE).

Use in a well-ventilated area.

FACESHIELD AND GLASSES   ACID GLOVES   APRON AND ARMGUARDS

Ammonium Chloride

[12125–02–9]

NH$_4$Cl

UNCLASSIFIED

HAZARDS

Dust High concentrations can cause irritation of skin, nose, throat, and lungs.

TLV/PEL 10 mg/m^3

STEL 20 mg/m^3

(Manufacturer)

PRECAUTIONS

Avoid skin and eye contact. Use acid or solvent gloves and safety glasses. Use in a well-ventilated area.

ACID GLOVES SAFETY GLASSES

Ammonium Citrate Dibasic [3012–65–5]

$(NH_4)_2HC_6H_5O_7$
Diammonium Citrate

UNCLASSIFIED

HAZARDS

Solid/Liquid   Can cause skin and eye irritation.

Dust/Mist   Can be mildly irritating to eyes and lungs.

(Manufacturer)

PRECAUTIONS

Avoid skin and eye contact. Use acid gloves and safety glasses.
Use with adequate ventilation.

ACID GLOVES   SAFETY GLASSES

Ammonium Dichromate [7789-09-5]

$(NH_4)_2Cr_2O_7$
Ammonium Bichromate

OXIDIZER

HAZARDS

Solid Can cause skin and eye irritation. Can cause skin ulceration (chrome sores).

Dust/Mist Can cause irritation of eyes, nose, throat, lungs, and skin. Can cause allergic skin reactions in some susceptible individuals. Can damage the nasal septum. Prolonged exposure can cause liver and kidney damage. Can cause liver cancer.

TLV/PEL 0.05 mg/m^3 (as water soluble, hexavalent chromium)

Important Hazard Suspected carcinogen.

OXIDIZER CANCER CAUSING

PRECAUTIONS

Avoid skin and eye contact. Use acid gloves, faceshield and safety glasses, apron and armguards.
Use only within the confines of an air-exhausted hood.

FACESHIELD AND GLASSES   ACID GLOVES   APRON AND ARMGUARDS   AIR EXHAUSTED

Ammonium Fluoride [12125-01-8]

NH$_4$F **CORROSIVE (ACID)**

HAZARDS

Solid/Liquid Can cause severe chemical burns to skin and eyes. Dilute solutions can cause skin rash.

Dust/Mist Irritating to skin, eyes, and lungs. High concentrations can cause chemical burns.

TLV/PEL 2.5 mg/m^3 (as fluoride)

CORROSIVE

PRECAUTIONS

Avoid skin and eye contact. Use acid gloves, faceshield and safety glasses, apron and armguards. Use in a well-ventilated area.
If decomposition occurs, handle as if hydrofluoric acid is present.

FACESHIELD AND GLASSES   ACID GLOVES   APRON AND ARMGUARDS

Ammonium Hydroxide

[1336–21–6]

NH_4OH

CORROSIVE (BASE)

HAZARDS

Liquid Can cause chemical burns to skin and eyes.

Vapor Characteristics similar to ammonia. Can be irritating to eyes, nose, and throat. High concentrations can cause fluid in lungs.

CORROSIVE

(Manufacturer)

PRECAUTIONS

Avoid skin and eye contact. Use acid gloves, faceshield and safety glasses, apron and armguards. Use only within the confines of an air-exhausted hood.

FACESHIELD AND GLASSES ACID GLOVES APRON AND ARMGUARDS AIR EXHAUSTED

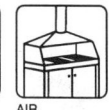

Ammonium Persulfate [7727–54–0]

$(NH_4)_2S_2O_8$ **OXIDIZER**
Ammonium Peroxydisulfate

HAZARDS

Solid Can cause chemical burns to skin and eyes. Prolonged skin contact may cause allergic skin reactions in some susceptible individuals.

Dust/Mist Can cause severe eye irritation. Can cause irritation of nose, throat, and lungs. May cause abdominal pain, nausea, and vomiting.

OXIDIZER

(Manufacturer)

PRECAUTIONS

Avoid skin and eye contact. Use acid gloves, faceshield and safety glasses, apron and armguards.
Use only within the confines of an air-exhausted hood.

FACESHIELD AND GLASSES   ACID GLOVES   APRON AND ARMGUARDS   AIR EXHAUSTED

Ammonium Phosphate

[7783–28–0]

$(NH_4)_3PO_4$

UNCLASSIFIED

HAZARDS

Solid/Liquid Can cause skin and eye irritation.

Dust/Mist Can be mildly irritating to eyes, nose, throat, and lungs.

(Manufacturer)

PRECAUTIONS

Avoid skin and eye contact. Use acid gloves and safety glasses.
Use with adequate ventilation.

ACID GLOVES   SAFETY GLASSES

Ammonium Sulfamate

[7773–06–0]

$NH_4SO_3NH_2$
Ammate

UNCLASSIFIED

HAZARDS

Solid Can cause mild skin and eye irritation.

Dust Can be mildly irritating to eyes and lungs.

TLV/PEL 10 mg/m^3 (total dust)
5 mg/m^3 (respirable fraction)

IDLH 1500 mg/m^3

Important Hazard Avoid mixing with acids as toxic fumes may be emitted.

SPECIAL HAZARD   OXIDIZER

(Manufacturer)

PRECAUTIONS

Avoid skin and eye contact. Use acid or solvent gloves and safety glasses.
Do not mix or store with acids.
Use in a well-ventilated area.

ACID GLOVES   SAFETY GLASSES

Antimony

[7440–36–0]

Sb

POISON

HAZARDS

Solid Can cause skin and eye irritation.

Dust Can cause severe irritation of skin, eyes, and lungs. Can cause stomach upset, pain or tightness in the chest, and shortness of breath. May cause damage to heart, liver, and kidneys.

TLV/PEL 0.5 mg/m^3

IDLH 50 mg/m^3

Important Hazard Reacts with acid to form a highly toxic gas, stibine (antimony hydride).

POISON   SPECIAL HAZARD

(Manufacturer)

PRECAUTIONS

Avoid skin and eye contact. Use acid or solvent gloves and safety glasses. Use only within the confines of an air-exhausted hood.

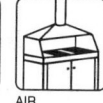

ACID GLOVES   SAFETY GLASSES   AIR EXHAUSTED

Antimony Trioxide

[1309-64-4]

Sb_2O_3

POISON

HAZARDS

Solid Can cause skin and eye irritation.

Dust Can cause severe irritation of skin, eyes, and lungs. Can cause stomach upset, pain or tightness in the chest, and shortness of breath. May cause damage to heart, liver, and kidneys.

TLV/PEL 0.5 mg/m^3 (as antimony)

Important Hazard Reacts with acid to form a highly toxic gas, stibine (antimony hydride). Suspected carcinogen.

| POISON | SPECIAL HAZARD | CANCER CAUSING | (Manufacturer) 2/0/0 |

PRECAUTIONS

Avoid skin and eye contact. Use acid or solvent gloves and safety glasses. Use only within the confines of an air-exhausted hood.

ACID GLOVES   SAFETY GLASSES   AIR EXHAUSTED

Aqua Regia

Approx. 80% HCl and 20% HNO_3　　　**CORROSIVE (ACID)**

HAZARDS

Liquid　Can cause severe burns to skin and eyes.

Vapor　Sharp chlorine odor detectable at low levels. Low concentrations can be irritating to eyes, nose, throat, and lungs. High concentrations can cause fluid in lungs and chemical burns to skin, eyes, lungs, and teeth.

TLV/PEL　See Hydrogen Chloride (HCl) and Nitric Acid (HNO_3)

Important Hazard　Emits highly toxic vapors when heated to decomposition. In contact with metals, flammable hydrogen gas may form.

| CORROSIVE | SPECIAL HAZARD | OXIDIZER | (Manufacturer) |

PRECAUTIONS

Avoid skin and eye contact. Use acid gloves, faceshield and safety glasses, apron and armguards. Use only within the confines of an air-exhausted hood. Avoid excessive heat and contact with metals.

| FACESHIELD AND GLASSES | ACID GLOVES | APRON AND ARMGUARDS | AIR EXHAUSTED |

Arsenic

[7440–38–2]

As **POISON**

HAZARDS

Solid Can be irritating to skin and eyes. Can cause discoloration of skin.

Dust Can be irritating to skin, eyes, and lungs. Can cause damage to nose and liver. May cause skin and lung cancer.

TLV/PEL 0.01 mg/m^3 **VP** 1 mm at 372°C

IDLH 5 mg/m^3

Important Hazard Suspected carcinogen. Reacts with acid to form a highly toxic gas, arsine.

POISON   CANCER CAUSING   SPECIAL HAZARD

(Manufacturer)

PRECAUTIONS

Use only within a closed, well-exhausted system or within the confines of an air-exhausted hood. Use acid or solvent gloves and safety glasses. Wash hands thoroughly after handling and before eating, drinking, or smoking.
The brushing, sweeping, or scraping of arsenic must be done in an air-exhausted hood or with local exhaust. Arsenic contaminated surfaces must be cleaned by dampening the area and wiping with damp paper towels; dry sweeping or brushing must be avoided. Portable vacuums with high-efficiency particulate filters, disposable coveralls, and respirators should be used for large accumulations. Do not use house vacuum systems.

ACID GLOVES   SAFETY GLASSES   AIR EXHAUSTED

Arsenic Trioxide

[1327–53–3]

As$_2$O$_3$
White Arsenic

POISON

HAZARDS

Dust Can be irritating to skin, eyes, and lungs. Can cause damage to nose and liver. May cause skin and lung cancer.

TLV/PEL 0.01 mg/m^3 (as arsenic)

IDLH 5 mg/m^3 (as arsenic)

Important Hazard Suspected carcinogen. Reacts with acid to form a highly toxic gas, arsine.

POISON | CANCER CAUSING | SPECIAL HAZARD

3 / 0 / 0

PRECAUTIONS

Arsenic trioxide is formed when arsine gas reacts with air over time or when trimethylarsenic decomposes.

Use only within a closed, well-exhausted system. Use acid or solvent gloves and safety glasses. Wash hands thoroughly after handling and before eating, drinking, or smoking.

Surfaces contaminated with arsenic trioxide must be cleaned by dampening the area and wiping with damp paper towels; dry sweeping or brushing must be avoided. Portable vacuums with high-efficiency particulate filters, disposable coveralls, and respirators should be used for large accumulations. Do not use house vacuum systems.

ACID GLOVES | SAFETY GLASSES | AIR EXHAUSTED

Arsine

[7784–42–1]

AsH$_3$

POISON/FLAMMABLE

HAZARDS

Gas Garlic odor detectable at 0.5 to 4 ppm. Can be irritating to eyes and nose. Can damage blood cells and cause dark red urine. Can cause nosebleed, dizziness, headache, nausea/vomiting, liver and kidney damage. Exposure to 250 ppm for 30 minutes may be fatal. Exposure to 500 ppm for a few minutes may be fatal. Flammable.

TLV/PEL 0.05 ppm

IDLH 3 ppm

ER 4 – 10%

Important Hazard Poisonous at low concentrations. Can be toxic at concentrations below the odor threshold. Suspected carcinogen.

POISON   COMBUSTIBLE   GAS HAZARD   SPECIAL HAZARD   CANCER CAUSING

PRECAUTIONS

Use only within a closed, well-exhausted system. Use only with a toxic gas monitor present and operating.

Keep away from ignition sources such as flame, spark, and heat. All suspected exposures must be reported to the medical department immediately.

SAFETY GLASSES   AIR EXHAUSTED   NO IGNITION SOURCE

Asbestos

[12001–29–5]
UNCLASSIFIED

HAZARDS

Fiber   Can be irritating to skin, eyes, and lungs. Can cause lung damage. Can cause lung cancer.

TLV/PEL   0.2 fibers/cc

Important Hazard   Confirmed human carcinogen.

CANCER CAUSING

(Manufacturer)

PRECAUTIONS

The brushing, sweeping, or scraping of asbestos must be done in an air-exhausted hood or with local exhaust. Asbestos-contaminated surfaces must be cleaned by dampening the area and wiping with damp paper towels, or a vacuum with a high-efficiency particulate filter. Do not disturb asbestos without first contacting your Environmental, Health, and Safety department. Use acid or solvent gloves and safety glasses. Depending on the work to be done, disposable coveralls and respirators may be needed.

Wash hands thoroughly after handling and before eating, drinking, or smoking. Local regulations may require additional precautions. Contact your safety representative for further guidance.

ACID GLOVES   SAFETY GLASSES   AIR EXHAUSTED

Barium Chloride

[10361–37–2]

BaCl$_2$ **POISON**

HAZARDS

Dust/Mist Can cause skin and nose irritation. High concentrations can cause lung damage. Higher concentrations may increase muscle excitability and cause paralysis. May cause changes to the blood and hemorrhages.

TLV/PEL 0.5 mg/m^3 (as barium)

IDLH 50 mg/m^3 (as barium)

POISON

(Manufacturer)

PRECAUTIONS

Avoid skin and eye contact. Use acid or solvent gloves and safety glasses. Use only within the confines of an air-exhausted hood.

ACID GLOVES   SAFETY GLASSES   AIR EXHAUSTED

Barium Hydroxide

[12230–71–6]

Ba(OH)$_2$

POISON

HAZARDS

Dust/Mist Can cause skin and nose irritation. High concentrations can cause lung damage. Higher concentrations may increase muscle excitability and cause paralysis. May cause changes to the blood and hemorrhages.

TLV/PEL 0.5 mg/m^3 (as barium)

POISON

(Manufacturer)

PRECAUTIONS

Avoid skin and eye contact. Use acid or solvent gloves and safety glasses. Use only within the confines of an air-exhausted hood.

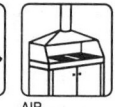

ACID GLOVES   SAFETY GLASSES   AIR EXHAUSTED

Barium Nitrate

[10022–31–8]

POISON/OXIDIZER

Ba(NO$_3$)$_2$
Barium Dinitrate

HAZARDS

Dust/Mist Can cause skin and nose irritation. High concentrations can cause lung damage. Higher concentrations may increase muscle excitability and cause paralysis. May cause changes to the blood and hemorrhages.

TLV/PEL 0.5 mg/m^3 (as barium)

IDLH 50 mg/m^3 (as barium)

POISON   OXIDIZER   (Manufacturer)

PRECAUTIONS

Avoid skin and eye contact. Use acid or solvent gloves and safety glasses. Use only within the confines of an air-exhausted hood.

ACID GLOVES   SAFETY GLASSES   AIR EXHAUSTED

Benzyl Alcohol

[100–51–6]

C₆H₅CH₂OH
Benzene Methanol

COMBUSTIBLE

HAZARDS

Liquid Can be readily absorbed through the skin. Can cause skin and eye irritation.

Vapor Aromatic odor detectable at 5.5 ppm. The low vapor pressure makes inhalation unlikely at room temperature, but high concentrations may cause irritation of the eyes, nose, and throat. High concentrations may also cause coughing, difficult breathing, dizziness, headache, dullness, tiredness, and may damage the blood. Combustible.

VP 0.15 mm at 25°C

FP 213°F

AT 817°F

PRECAUTIONS

Particular care should be taken to avoid skin and eye contact. Use solvent gloves, faceshield and safety glasses, apron and armguards. Use only within the confines of an air-exhausted hood.

Keep away from open flame and high temperatures.

Beryllium

[7440–41–7]

Be **POISON**

HAZARDS

Dust Small amounts of dust can cause eye, skin, and lung irritation. High concentrations can cause fluid in the lungs and damage to the lungs. Repeated skin contact can cause a skin rash allergy. May cause lung cancer. Combustible under certain circumstances.

TLV/PEL 0.002 mg/m^3 (0.025 mg/m^3 C)

IDLH 4 mg/m^3

STEL 0.005 mg/m^3 (30 minute average)

Important Hazard Emits very toxic fumes of beryllium oxide when heated to decomposition. Suspected carcinogen.

CANCER CAUSING | SPECIAL HAZARD

3 / 1 / 0

PRECAUTIONS

Particular care must be taken to avoid inhalation and skin and eye contact. Use acid or solvent gloves and safety glasses. Use same precautions as for arsenic.

Use within a closed, air-exhausted system if the possibility of dust exists.

ACID GLOVES | SAFETY GLASSES | AIR EXHAUSTED

Beryllium Oxide

[1304–56–9]

BeO
Beryllia

POISON

HAZARDS

Dust Small amounts of dust can cause skin, eye, and lung irritation. High inhalation exposures can cause fluid in the lungs and damage to the lungs. Repeated skin contact can cause a skin rash allergy. May cause lung cancer. Combustible under certain circumstances.

TLV/PEL 0.002 mg/m^3 (0.025 mg/m^3 C)

IDLH 4 mg/m^3 (as beryllium)

STEL 0.005 mg/m^3 (30 minute average)

Important Hazard Suspected carcinogen.

CANCER CAUSING

PRECAUTIONS

Particular care must be taken to avoid inhalation and eye and skin contact. Use acid or solvent gloves and safety glasses. Use same precautions as for arsenic.

Use within a closed, air-exhausted system if the possibility of dust exists.

ACID GLOVES   SAFETY GLASSES   AIR EXHAUSTED

Boric Acid

H$_3$BO$_3$

[10043–35–3]

**UNCLASSIFIED
(WEAK ACID)**

HAZARDS

Dust Can cause mild skin and eye irritation. In high concentrations can cause nausea, vomiting, and diarrhea.

0 / 0 / 0 (Manufacturer)

PRECAUTIONS

Avoid skin and eye contact. Use acid gloves and safety glasses.
Use in a well-ventilated area.

ACID GLOVES   SAFETY GLASSES

Boron Nitride

[10043–11–5]

BN

UNCLASSIFIED

HAZARDS

Dust Nuisance dust. Can cause slight eye, nose, and throat irritation.

TLV/PEL 10 mg/m^3 (as a nuisance dust)

PRECAUTIONS

Avoid skin and eye contact. Use acid or solvent gloves and safety glasses. Use in a well-ventilated area.

SAFETY GLASSES

ACID GLOVES

Boron Oxide

[1303–86–2]

B_2O_3
Boron Trioxide
Boric Anhydride

UNCLASSIFIED

HAZARDS

Solid Can cause mild skin and eye irritation.

Dust Can be mildly irritating to eyes. Can cause dryness of the mouth, nose, and throat, and irritation of the throat and cough.

TLV/PEL 10 mg/m^3

IDLH 2000 mg/m^3

(Manufacturer)

PRECAUTIONS

Avoid skin and eye contact. Use acid gloves and safety glasses.
Use in a well-ventilated area. Keep away from water, including small amounts of moisture.

SAFETY GLASSES   ACID GLOVES   WATER REACTIVE

Boron Tribromide

[10294–33–4]

BBr$_3$

CORROSIVE (ACID)

HAZARDS

Liquid Can cause severe burns to skin and eyes.

Vapor Can be irritating to skin, eyes, and lungs.

TLV/PEL 1 ppm C

VP 40 mm at 14°C

Important Hazard Forms hydrogen bromide (HBr) gas on contact with water or moisture in the air. Can explode when heated.

CORROSIVE	SPECIAL HAZARD	GAS HAZARD

NFPA: Health 2, Flammability 0, Reactivity 3, W

PRECAUTIONS

Avoid skin and eye contact. Use acid gloves, faceshield and safety glasses, apron and armguards. Use only within the confines of an air-exhausted hood. Keep away from heat.
Keep away from water, including small amounts of moisture.

FACESHIELD AND GLASSES	ACID GLOVES	APRON AND ARMGUARDS	AIR EXHAUSTED	WATER REACTIVE

147

Boron Trichloride

[10294–34–5]

BCl_3

CORROSIVE (ACID)

HAZARDS

Gas Can cause severe skin, eye, nose, throat, and lung irritation.

Important Hazard Forms hydrogen chloride (HCl) gas on contact with water or moisture in the air (see Hazards of Hydrogen Chloride).

CORROSIVE   GAS HAZARD   SPECIAL HAZARD

(Manufacturer)

PRECAUTIONS

Avoid skin and eye exposure and inhalation. Use only within a closed system.

Keep away from water, including small amounts of moisture.

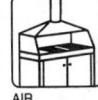

SAFETY GLASSES   AIR EXHAUSTED   WATER REACTIVE

Boron Trifluoride

[7637–07–2]

BF_3

CORROSIVE (ACID)

HAZARDS

Gas Can cause severe skin, eye, nose, throat, and lung irritation. Can cause fluid in the lungs.

TLV/PEL 1 ppm C

IDLH 25 ppm

Important Hazard Forms hydrogen fluoride (HF) gas on contact with water or moisture in the air. Skin and eye contact is not immediately evident; pain may not start for up to 24 hours (see Hazards of Hydrogen Fluoride).

CORROSIVE   GAS HAZARD   SPECIAL HAZARD

0 / 4 / 1

PRECAUTIONS

Avoid skin and eye exposure and inhalation. Use only within a closed system.
Keep away from water, including small amounts of moisture.

SAFETY GLASSES   AIR EXHAUSTED   WATER REACTIVE

Butane

[106–97–8]

C_4H_4

FLAMMABLE

HAZARDS

Gas Odor detectable at 2600 to 5000 ppm. High concentrations (10,000 ppm for 10 minutes) can cause dizziness. Flammable.

TLV/PEL	800 ppm
FP	–76°F
AT	761°F
ER	1.9 – 8.5%

FLAMMABLE

4/1/0

PRECAUTIONS

Use in a well-ventilated area.
Keep away from ignition sources such as flame, spark, and heat.

NO IGNITION SOURCE

2-Butoxyethanol

[111–76–2]

C$_4$H$_9$OCH$_2$CH$_2$OH
Butyl Cellosolve

COMBUSTIBLE

HAZARDS

Liquid Can be readily absorbed through the skin. Repeated contact can cause skin irritation.

Vapor Sweet, rancid odor detectable at 0.1 to 0.5 ppm. Can cause eye, nose, and throat irritation. Can cause blood cell damage. May cause lung, kidney, and liver damage. High concentrations can cause headache, dizziness, drowsiness, and difficulty in breathing. Combustible.

TLV/PEL	25 ppm	**FP**	160°F
IDLH	700 ppm	**ER**	1.1 – 10.6%
VP	0.6 mm	**AT**	472°F

Important Hazard Suspected of causing reproductive disorders.

COMBUSTIBLE   SPECIAL HAZARD

2 / 2 / 0

PRECAUTIONS

Particular care should be taken to avoid skin and eye contact. Use solvent gloves, faceshield and safety glasses, apron and armguards. Use only within the confines of an air-exhausted hood.

Keep away from open flame and high temperatures.

FACESHIELD AND GLASSES   SOLVENT GLOVES   APRON AND ARMGUARDS   AIR EXHAUSTED   NO IGNITION SOURCE

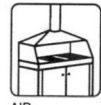

2-(2-Butoxyethoxy) Ethanol

[112–34–5]

Diethylene Glycol Monobutyl Ether
Butyl Carbitol

COMBUSTIBLE

HAZARDS

Liquid Can be readily absorbed through the skin. Repeated contact can cause skin and eye irritation.

Vapor/Mist If heated, or a mist is formed, can cause concentrations high enough to cause headache, nausea, dizziness, drowsiness, incoordination, and eye irritation. This chemical is not likely to form significant vapor concentrations at room temperature (26 ppm is the maximum concentration at room temperature in a confined space). Combustible.

VP 0.02 mm

AT 442°F

FP 172°F

Important Hazard Suspected of causing reproductive disorders.

COMBUSTIBLE   SPECIAL HAZARD

PRECAUTIONS

Particular care should be taken to avoid skin and eye contact. Use solvent gloves, faceshield and safety glasses, apron and armguards. Use only within the confines of an air-exhausted hood.

Keep away from open flame and high temperatures.

FACESHIELD AND GLASSES   SOLVENT GLOVES   APRON AND ARMGUARDS   AIR EXHAUSTED   NO IGNITION SOURCE

2-Butoxyethyl Acetate

[112–07–2]

Ethylene Glycol Monobutyl Ether Acetate

COMBUSTIBLE

HAZARDS

Liquid Repeated contact can cause skin and eye irritation. Can be readily absorbed through the skin.

Vapor Fruity odor detectable at 0.1 to 0.4 ppm. Can be irritating to skin, eyes, and lungs. May cause lung, blood, kidney, and liver changes. High concentrations can cause headache, dizziness, drowsiness, and difficulty in breathing. Combustible.

TLV/PEL   20 ppm (German)

VP   0.3 mm

FP   160°F

AT   645°F

Important Hazard   Suspected of causing reproductive disorders.

COMBUSTIBLE   SPECIAL HAZARD

(1 / 2 / 0)

PRECAUTIONS

Particular care should be taken to avoid skin and eye contact. Use solvent gloves, faceshield and safety glasses, apron and armguards. Use only within the confines of an air-exhausted hood.

Keep away from open flame and high temperatures.

FACESHIELD AND GLASSES   SOLVENT GLOVES   APRON AND ARMGUARDS   AIR EXHAUSTED   NO IGNITION SOURCE

n-Butyl Acetate

[123–86–4]

CH$_3$COO(CH$_2$)$_3$CH$_3$
NBA

FLAMMABLE

HAZARDS

Liquid Can cause eye irritation. Repeated contact can cause skin irritation.

Vapor Sweet, fruity odor detectable at 0.06 to 7 ppm. Can cause irritation of nose and throat at 200 to 300 ppm. High concentrations can cause headache and dizziness. Flammable.

TLV/PEL	20 ppm	**FP**	72°F
IDLH	1700 ppm	**AT**	797°F
VP	10 mm	**ER**	1.3 – 7.6%

FLAMMABLE

PRECAUTIONS

Avoid skin and eye contact. Use solvent gloves, faceshield and safety glasses, apron and armguards. Use only within the confines of an air-exhausted hood.

Keep away from ignition sources such as flame, spark, and heat.

FACESHIELD AND GLASSES   SOLVENT GLOVES   APRON AND ARMGUARDS   AIR EXHAUSTED   NO IGNITION SOURCE

n-Butyl Alcohol

[71–36–3]

C_4H_9OH
n-Butanol

FLAMMABLE

HAZARDS

Liquid Repeated contact can cause skin and eye irritation. Can be readily absorbed through the skin.

Vapor Odor detectable at 0.1 to 10 ppm. Can be irritating to nose and throat at 24 ppm. Can cause headaches and eye irritation at 50 ppm. At higher concentrations can cause hearing loss and dizziness. Flammable.

TLV/PEL	50 ppm C	**FP**	84°F
IDLH	1400 ppm	**AT**	650°F
VP	4.2 mm	**ER**	1.4 – 11%

FLAMMABLE

PRECAUTIONS

Particular care should be taken to avoid skin and eye contact. Use solvent gloves, faceshield and safety glasses, apron and armguards. Use only within the confines of an air-exhausted hood.

Keep away from ignition sources such as flame, spark, and heat.

FACESHIELD AND GLASSES   SOLVENT GLOVES   APRON AND ARMGUARDS   AIR EXHAUSTED   NO IGNITION SOURCE

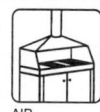

Butyl Glycidyl Ether

[2426–08–6]

COMBUSTIBLE

$C_4H_9OCH_2CHOCH_2$
BGE
Epoxy Butoxypropane

HAZARDS

Liquid Can cause skin and eye irritation. Repeated contact can cause an allergic skin rash.

Vapor Irritating but not unpleasant odor. May be irritating to eyes, nose, throat, and lungs. High concentrations may cause dizziness and loss of appetite. Combustible.

TLV/PEL 25 ppm

IDLH 250 ppm

VP 3 mm

FP 130°F

COMBUSTIBLE

PRECAUTIONS

Avoid skin and eye contact. Use solvent gloves, faceshield and safety glasses, apron and armguards. Use only within the confines of an air-exhausted hood.
Keep away from open flame and high temperatures.

FACESHIELD AND GLASSES | SOLVENT GLOVES | APRON AND ARMGUARDS | AIR EXHAUSTED | NO IGNITION SOURCE

Butyrolactone

[96–48–0]

OCH$_2$CH$_2$CH$_2$CO

COMBUSTIBLE

HAZARDS

Liquid May cause skin and eye irritation.

Vapor Mild odor. High concentrations may cause dizziness. Combustible. When heated to decomposition, it emits acrid and irritating fumes.

VP < 1 mm

FP 209°F

ER 3.6 – 16%

Important Hazard Suspected carcinogen.

COMBUSTIBLE CANCER CAUSING

PRECAUTIONS

Avoid skin and eye contact. Use solvent gloves, faceshield and safety glasses, apron and armguards. Use only within the confines of an air-exhausted hood.

Keep away from open flame and high temperatures.

FACESHIELD AND GLASSES SOLVENT GLOVES APRON AND ARMGUARDS AIR EXHAUSTED NO IGNITION SOURCE

Calcium Carbonate

[1317-65-3]

CaCO$_3$
Limestone

UNCLASSIFIED

HAZARDS

Dust Nuisance dust. Can cause slight eye, nose, and throat irritation. High concentrations may cause sneezing and coughing.

TLV/PEL 10 mg/m^3

(Manufacturer)

PRECAUTIONS

Avoid skin and eye contact. Use acid or solvent gloves and safety glasses. Use in a well-ventilated area.

SAFETY GLASSES   ACID GLOVES

Calcium Chloride

[10043–52–4]

CaCl$_2$

UNCLASSIFIED

HAZARDS

Dust Can cause skin, eyes, nose, and throat irritation.

<!-- NFPA diamond: Health 3, Fire 1, Reactivity 0 -->
(Manufacturer)

PRECAUTIONS

Avoid skin and eye contact. Use acid or solvent gloves and safety glasses. Use in a well-ventilated area.

SAFETY GLASSES   ACID GLOVES

Calcium Hypochlorite

[7778–54–3]

$Ca(OCl)_2$
Calcium Oxychloride

OXIDIZER

HAZARDS

Dust May have strong chlorine odor caused by decomposition. Can cause skin, eye, nose and throat irritation.

OXIDIZER

PRECAUTIONS

Avoid skin and eye contact. Use acid gloves, faceshield and safety glasses, apron and armguards.

Use in a well-ventilated area.

FACESHIELD AND GLASSES   ACID GLOVES   APRON AND ARMGUARDS

Calcium Oxide

CaO
Quicklime

[1305–78–8]
UNCLASSIFIED

HAZARDS

Dust Can be very irritative to skin, eyes, nose, and throat.
TLV/PEL 2 mg/m^3
IDLH 25 mg/m^3

3 / 0 / 1

PRECAUTIONS

Avoid skin and eye contact. Use acid or solvent gloves and safety glasses.
Use only in the confines of an air-exhausted hood.

SAFETY GLASSES   ACID GLOVES   AIR EXHAUSTED

Calcium Sulfate

[7778–18–9]

$CaSO_4$ **UNCLASSIFIED**

HAZARDS

Dust Nuisance dust. Can cause slight eye, nose, and throat irritation.

TLV/PEL 10 mg/m^3

(Manufacturer)

PRECAUTIONS

Avoid skin and eye contact. Use acid or solvent gloves and safety glasses. Use in a well-ventilated area.

SAFETY GLASSES

ACID GLOVES

Carbon Black

C
Thermal Black

[1333–86–4]

UNCLASSIFIED

HAZARDS

Dust Can be slightly irritating to skin, eyes, nose, and throat.

TLV/PEL 3.5 mg/m^3

IDLH 1750 mg/m^3

Important Hazard Suspected carcinogen.

CANCER CAUSING

(Manufacturer)

PRECAUTIONS

Avoid inhalation and skin and eye contact. Use acid or solvent gloves and safety glasses.

Use in a well-ventilated area or within the confines of an air-exhausted hood.

SAFETY GLASSES   ACID GLOVES   AIR EXHAUSTED

Carbon Dioxide

[124–38–9]

CO_2

NONFLAMMABLE GAS

HAZARDS

Solid Solid CO_2 or "dry ice" freezes skin on contact.

Gas Can displace oxygen in air causing headache and dizziness at very high concentrations (acts as a simple asphyxiant).

TLV/PEL 5000 ppm

STEL 30,000 ppm

IDLH 40,000 ppm

PRECAUTIONS

Use heavy leather, cotton, or cryogenic gloves when handling dry ice.
Use in a well-ventilated area.

SAFETY GLASSES

Carbon Disulfide

[75–15–0]

CS_2
Carbon Bisulfide

FLAMMABLE

HAZARDS

Liquid Can be readily absorbed through the skin. Can cause skin and eye irritation. Repeated contact can cause skin rash.

Vapor Slightly pungent vegetable-like odor detectable at 0.02 to 0.4 ppm. Exposures at 500 to 1000 ppm may cause severe mood and personality disturbances. Exposure to 4800 ppm for 30 minutes can cause coma and may be fatal. Repeated exposure can cause damage to the heart and liver, and vision and stomach problems. Flammable.

TLV/PEL	10 ppm	**FP**	–22°F
IDLH	500 ppm	**AT**	212°F
VP	300 mm	**ER**	1 – 50%

Important Hazard Vapors are extremely flammable.

FLAMMABLE   SPECIAL HAZARD

4 / 3 / 0

PRECAUTIONS

Avoid skin and eye contact. Use solvent gloves, faceshield and safety glasses, apron and armguards. Use only in the confines of an air-exhausted hood.

Keep away from ignition sources such as flame, spark, and heat.

FACESHIELD AND GLASSES   SOLVENT GLOVES   APRON AND ARMGUARDS   AIR EXHAUSTED   NO IGNITION SOURCE

Carbon Monoxide

[630–08–0]

CO

FLAMMABLE

HAZARDS

Gas Displaces oxygen in the blood. Exposures at 50 to 400 ppm may cause headache. Above 400 ppm may cause weakness, dizziness, nausea, and fainting. Above 1200 ppm increased heartbeat, irregular heartbeat. Above 2000 ppm loss of consciousness and death. Highly flammable.

TLV/PEL 25 ppm (200 ppm C)

IDLH 1200 ppm

AT 1125°F

ER 12 – 74%

Important Hazard Carbon monoxide does not have an odor or cause irritation at toxic or lethal concentrations.

| FLAMMABLE | GAS HAZARD | SPECIAL HAZARD |

PRECAUTIONS

Use in a closed system or within the confines of an air-exhausted hood. Keep away from ignition sources such as flame, spark, and heat.

| SAFETY GLASSES | AIR EXHAUSTED | NO IGNITION SOURCE |

Carbon Tetrachloride [56–23–5]

CCl$_4$ **UNCLASSIFIED**
Tetrachloromethane

HAZARDS

Liquid Can be readily absorbed through the skin. Repeated contact can cause skin and eye irritation.

Vapor Sweet, pungent odor detectable at 140 to 580 ppm. Concentrations of 25 to 30 ppm can cause nausea, vomiting, dizziness, drowsiness, and headache. Long-term exposure to concentrations at or above 200 ppm may cause liver and kidney damage. Exposure to alcohol and carbon tetrachloride may be particularly toxic.

TLV/PEL	2 ppm	**STEL**	10 ppm
IDLH	200 ppm	**VP**	91 mm

Important Hazard Suspected carcinogen. May be toxic at concentrations below the odor threshold.

CANCER CAUSING   SPECIAL HAZARD

PRECAUTIONS

Particular care should be taken to avoid skin and eye contact. Use solvent gloves, faceshield and safety glasses, apron and armguards. Use only within the confines of an air-exhausted hood or closed system.

Particular care should be taken to avoid contact with the liquid and inhalation of vapors.

FACESHIELD AND GLASSES   SOLVENT GLOVES   APRON AND ARMGUARDS   AIR EXHAUSTED

Carbon Tetrafluoride [75–73–0]

CF_4
Tetrafluoromethane
Freon-14

NONFLAMMABLE GAS

HAZARDS

Gas Odorless gas. Extremely high concentrations (i.e., greater than 40,000 ppm) can cause dizziness and drowsiness. May also cause heart irregularities.

(Manufacturer)

PRECAUTIONS

Use in a well-ventilated area.

Note: This compound contributes to global warming if released to the atmosphere.

SAFETY GLASSES

Catechol

[120–80–9]

$C_6H_4(OH)_2$
Pyrocatechol

**POISON/
CORROSIVE(ACID)/
COMBUSTIBLE**

HAZARDS

Solid/Liquid Can be readily absorbed through the skin. Can cause skin and eye irritation. High concentrations can cause convulsions, blood damage, and kidney damage.

Vapor Can cause eye, nose, throat, and lung irritation. High concentrations may cause convulsions. Combustible.

TLV/PEL 5 ppm

VP 10 mm at 118.3°C

FP 261°F

Important Hazard Liquid can be absorbed through the skin in amounts capable of causing death.

POISON CORROSIVE COMBUSTIBLE SPECIAL HAZARD

(Manufacturer) 1/2/0

PRECAUTIONS

Particular care should be taken to avoid skin contact with catechol.
Avoid all skin and eye contact. Use solvent gloves, faceshield and safety glasses, apron and armguards. Use only within the confines of an air-exhausted hood. Keep away from open flame and high temperatures.

FACESHIELD AND GLASSES SOLVENT GLOVES APRON AND ARMGUARDS AIR EXHAUSTED NO IGNITION SOURCE

Chlorine

[7782–50–5]

Cl_2

POISON/OXIDIZER

HAZARDS

Gas Greenish yellow gas with strong, irritating odor detectable at 0.03 to 0.4 ppm. Can be highly irritating to skin, eyes, and lungs. Mild mucous membrane irritation can occur at 0.2 to 16 ppm; eye irritation occurs at 7 to 8 ppm; throat irritation at 15 ppm; and cough at 30 ppm. Very high concentrations can cause fluid in the lungs. A few deep breaths at 1000 ppm can cause death.

TLV/PEL 0.5 ppm (1 ppm C)

IDLH 10 ppm

Important Hazard Poisonous if inhaled. Can be fatal at high concentrations. Oxidizer.

| POISON | GAS HAZARD | OXIDIZER | SPECIAL HAZARD |

4 / 0 / 0 / OX

PRECAUTIONS

Use only within a closed, well-exhausted system. Avoid all direct contact. Keep away from other chemicals.

SAFETY GLASSES — AIR EXHAUSTED

170

Chlorobenzene

[108–90–7]

C_6H_5Cl
Monochlorobenzene

FLAMMABLE

HAZARDS

Liquid Can cause skin and eye irritation. Repeated contact can cause skin rash.

Vapor Almond-like odor detectable at 0.2 to 0.5 ppm. High concentrations can cause eye, nose, and skin irritation and intoxication. Eye and nose irritation occurs at 200 ppm. Prolonged exposure may cause headache; dizziness; digestive disorders; lung, liver, and kidney damage; numbness of hands and feet. Flammable.

TLV/PEL	10 ppm	**FP**	85°F
IDLH	1000 ppm	**AT**	1180°F
VP	8.8 mm	**ER**	1.3 – 7.1%

FLAMMABLE

PRECAUTIONS

Avoid skin and eye contact. Use solvent gloves, faceshield and safety glasses, apron and armguards. Use only in the confines of an air-exhausted hood.

Keep away from ignition sources such as flame, spark, and heat.

FACESHIELD AND GLASSES   SOLVENT GLOVES   APRON AND ARMGUARDS   AIR EXHAUSTED   NO IGNITION SOURCE

Chloroform

[67–66–3]

CHCl$_3$
Trichloromethane

UNCLASSIFIED

HAZARDS

Liquid Can cause skin and eye irritation. May be readily absorbed through the skin.

Vapor Sweet odor detectable at 200 to 300 ppm. Exposure at about 1000 ppm may cause dizziness, headache, fatigue, and slight shortness of breath after a few minutes. 4000 ppm may cause vomiting and a feeling of fainting. Prolonged exposure may cause fatigue, digestive disturbances, frequent burning urination, mental dullness, and enlargement of the liver. May cause liver cancer.

TLV/PEL 2 ppm

IDLH 500 ppm

VP 160 mm

Important Hazard Suspected carcinogen. May be toxic at concentrations below the odor threshold.

CANCER CAUSING   SPECIAL HAZARD

PRECAUTIONS

Avoid skin and eye contact. Use solvent gloves, faceshield and safety glasses, apron and armguards. Use only within the confines of an air-exhausted hood or closed system.

FACESHIELD AND GLASSES   SOLVENT GLOVES   APRON AND ARMGUARDS   AIR EXHAUSTED

o-Chlorotoluene

[95–49–8]

C_7H_7Cl

COMBUSTIBLE

HAZARDS

Liquid Can cause skin and eye irritation.

Vapor High concentrations can cause eye, nose, and skin irritation and intoxication. Combustible.

TLV/PEL 50 ppm

VP 2.7 mm

FP 117°F

COMBUSTIBLE

2 / 2 / 0

PRECAUTIONS

Avoid skin and eye contact. Use solvent gloves, faceshield and safety glasses, apron and armguards. Use only in the confines of an air-exhausted hood.

Keep away from open flame and high temperatures.

FACESHIELD AND GLASSES | SOLVENT GLOVES | APRON AND ARMGUARDS | AIR EXHAUSTED | NO IGNITION SOURCE

Chromic Acid

[1333–82–0]

CrO$_3$
Chromium Trioxide

**OXIDIZER/
CORROSIVE (ACID)**

HAZARDS

Solid/Liquid Can cause chemical burns to skin and eyes.

Dust/Mist 0.06 mg/m^3 can cause nose irritation. 0.11 to 0.15 mg/m^3 can cause damage to nose and irritation of eyes, nose, and throat. Can cause asthmatic bronchitis and liver and kidney damage.

TLV/PEL 0.05 mg/m^3 (0.1 mg/m^3 C)

IDLH 15 mg/m^3

Important Hazard Suspected human carcinogen.

OXIDIZER   CORROSIVE   CANCER CAUSING

PRECAUTIONS

Avoid all skin and eye contact. Use acid gloves, faceshield and safety glasses, apron and armguards. Use only in the confines of an air-exhausted hood.
All slow-healing skin sores must be reported to the medical department.

FACESHIELD AND GLASSES   ACID GLOVES   APRON AND ARMGUARDS   AIR EXHAUSTED

Chromic Sulfate

[10031-37-5]

UNCLASSIFIED

$Cr_2(SO_4)_3$
Chromium Sulfate
Chromous Salts

HAZARDS

Dust/Mist Can cause irritation of skin, eyes, and lungs. Can cause lung damage.

TLV/PEL 0.5 mg/m^3 (as Cr III compound)

IDLH 250 mg/m^3

PRECAUTIONS

Avoid skin and eye contact. Use acid or solvent gloves and safety glasses. Use in a well-ventilated area.

ACID GLOVES

SAFETY GLASSES

Chromium, Hexavalent

Cr VI **UNCLASSIFIED**

HAZARDS

Solid/Liquid Can cause irritation of skin and eyes. Can cause an allergic skin rash.

Dust/Mist Can cause severe irritation of nose, lungs, skin, and eyes. 0.11 to 0.15 mg/m^3 can cause ulcers of the nasal septum and irritation of the eyes, nose, and throat. Can cause asthmatic bronchitis and liver and kidney damage. Water-insoluble forms can cause lung cancer.

TLV/PEL 0.01 mg/m^3

IDLH 15 mg/m^3 (water-soluble forms)

Important Hazard Confirmed human carcinogen (water-insoluble forms).

CANCER CAUSING

PRECAUTIONS

Avoid all skin and eye contact. Use acid gloves, faceshield and safety glasses, apron and armguards. Use only in the confines of an air-exhausted hood.

Particular care should be taken to avoid inhalation. All slow-healing skin lesions (sores) must be reported to the medical department.

FACESHIELD AND GLASSES   ACID GLOVES   APRON AND ARMGUARDS   AIR EXHAUSTED

Citric Acid

[77–92–9]

$C_6H_8O_7$

CORROSIVE (ACID)

HAZARDS

Liquid Can cause irritation of skin and severe irritation and burns of the eyes. Can cause allergic skin reaction.

Mist May cause irritation of the nose, throat, and skin. Higher concentrations may cause coughing, sneezing, and difficulty breathing.

AT 1850°F

CORROSIVE

(Manufacturer)

PRECAUTIONS

Avoid skin and eye contact. Use acid gloves, faceshield and safety glasses, apron and armguards.

Use in a well-ventilated area.

FACESHIELD AND GLASSES   ACID GLOVES   APRON AND ARMGUARDS

Cobalt Nitrate

[10141-05-6]

$Co(NO_3)_2$
Cobaltous Nitrate

OXIDIZER

HAZARDS

Dust/Mist Can cause irritation of eyes, lungs, and skin. Can cause changes in blood cells.

TLV/PEL 0.02 mg/m^3 (as cobalt)

Important Hazard Suspected carcinogen.

OXIDIZER   CANCER CAUSING

PRECAUTIONS

Avoid skin and eye contact. Use acid or solvent gloves, faceshield and safety glasses, apron and armguards.
Use only within the confines of an air-exhausted hood.

FACESHIELD AND GLASSES   ACID GLOVES   APRON AND ARMGUARDS   AIR EXHAUSTED

Cobalt Sulfamate [10124–43–3]

$Co(NH_2SO_3)_2$ **UNCLASSIFIED**

HAZARDS

Dust/Mist Can cause irritation of eyes, lungs, and skin. May cause an allergic response.

TLV/PEL 0.02 mg/m^3 (as cobalt)

Important Hazard Suspected carcinogen.

CANCER CAUSING

PRECAUTIONS

Avoid skin and eye contact. Use acid or solvent gloves, faceshield and safety glasses, apron and armguards.
Use only within the confines of an air-exhausted hood.

FACESHIELD AND GLASSES   ACID GLOVES   APRON AND ARMGUARDS   AIR EXHAUSTED

Copper

[7440-50-8]

Cu

UNCLASSIFIED

HAZARDS

Dust/Mist/Fume Can cause irritation of skin, eyes, nose, and lungs. Can cause metallic or sweet taste, nausea, metal fume fever, and in some instances discoloration of the skin and hair. Welding with copper can cause metal fume fever.

TLV/PEL 0.2 mg/m^3 (fume)
1 mg/m^3 (dust/mist)

IDLH 100 mg/m^3

(Manufacturer)

PRECAUTIONS

Avoid skin and eye contact. Use acid or solvent gloves and safety glasses. Use in a well-ventilated area.

ACID GLOVES   SAFETY GLASSES

Copper Chloride

[1344–67–8]

CuCl$_2$
Cupric Chloride

UNCLASSIFIED

HAZARDS

Dust/Mist Can cause coughing and irritation of nose, throat, and lungs. Skin and eye contact can cause irritation. Weak acid.

TLV/PEL 1 mg/m^3 (as copper)

(Manufacturer)

PRECAUTIONS

Avoid skin and eye contact. Use acid or solvent gloves, faceshield and safety glasses, apron and armguards.
Use in a well-ventilated area.

FACESHIELD AND GLASSES — ACID GLOVES — APRON AND ARMGUARDS

Copper Nitrate

[3251–23–8]

Cu(NO$_3$)$_2$
Cupric Nitrate

OXIDIZER

HAZARDS

Dust/Mist Can cause irritation of skin, eyes, nose, throat, and lungs. High concentrations may cause damage to blood cells, nervous system, and kidneys.

TLV/PEL 1 mg/m^3 (as copper)

OXIDIZER

(Manufacturer)

PRECAUTIONS

Avoid skin and eye contact. Use acid or solvent gloves, faceshield and safety glasses, apron and armguards.
Use in a well-ventilated area.

FACESHIELD AND GLASSES   ACID GLOVES   APRON AND ARMGUARDS

Copper Oxide

[1317–39–1]

CuO
Cupric Oxide

UNCLASSIFIED

HAZARDS

Dust/Mist Can cause irritation of skin, eyes, nose, and lungs. High concentrations may cause damage to blood cells, nervous system, and kidneys.

TLV/PEL 1 mg/m^3 (as copper dust or mist)
0.2 mg/m^3 (as copper fume)

IDLH 100 mg/m^3 (as copper)

(Manufacturer) 2 0 0

PRECAUTIONS

Avoid skin and eye contact. Use acid or solvent gloves, faceshield and safety glasses, apron and armguards.
Use in a well-ventilated area.

FACESHIELD AND GLASSES   ACID GLOVES   APRON AND ARMGUARDS

Copper Sulfate

[7758-98-7]

$CuSO_4$
Cupric Sulfate

UNCLASSIFIED

HAZARDS

Dust/Mist Can cause irritation of skin, eyes, nose, and throat. Prolonged exposure to mists may cause weakness, loss of appetite and weight, and cough. Some individuals may become sensitized from repeat or prolonged exposure and may develop allergic skin rash.

TLV/PEL 1 mg/m^3 (as copper)

PRECAUTIONS

Avoid skin and eye contact. Use acid or solvent gloves, faceshield and safety glasses, apron and armguards.
Use in a well-ventilated area.

FACESHIELD AND GLASSES	ACID GLOVES	APRON AND ARMGUARDS

Cresol

[1319–77–3]

$CH_3C_6H_4OH$

COMBUSTIBLE

HAZARDS

Liquid Creosote/phenol-like odor detectable at 0.00005 to 0.008 ppm. Can be readily absorbed through the skin. Can cause skin and eye irritation. High concentrations may cause convulsions, blood damage, and kidney damage.

Vapor Can cause eye and lung irritation. Combustible.

TLV/PEL	5 ppm	**AT**	1036 – 1110°F
IDLH	250 ppm		
VP	0.11 – 0.25 mm		
FP	178 – 187°F		

Important Hazard Liquid can be absorbed through the skin in amounts capable of causing death.

POISON   CORROSIVE   COMBUSTIBLE   SPECIAL HAZARD

2 / 3 / 0

PRECAUTIONS

Particular care should be taken to avoid skin contact with cresol.

Avoid all skin and eye contact. Use solvent gloves, faceshield and safety glasses, apron and armguards. Use only within the confines of an air-exhausted hood.

FACESHIELD AND GLASSES   SOLVENT GLOVES   APRON AND ARMGUARDS   AIR EXHAUSTED

185

Cyanide Salts

NaCN, KCN, AuCN, AgCN, CuCN

POISON/ CORROSIVE (BASE)

HAZARDS

Solid Can be readily absorbed through the skin. Repeated contact can cause skin and eye irritation.

Dust Cyanide salts are odorless. Can be irritating to eyes and lungs. Reacts with acids to form hydrogen cyanide gas (HCN). HCN has an almond-like odor (see Hydrogen Cyanide).

TLV/PEL 5 mg/m^3 C **IDLH** 25 mg/m^3 (as cyanide)

Important Hazard Cyanide salts can be readily absorbed through the skin. They can react with acidic chemicals to form hydrogen cyanide — a highly poisonous gas. Asphyxia and death can occur from high exposure levels. 270 ppm of hydrogen cyanide is immediately fatal.

POISON   CORROSIVE   SPECIAL HAZARD

(as NaCN)

PRECAUTIONS

Avoid skin and eye contact. Use acid gloves, apron and armguards, and safety glasses. Use only within the confines of an air-exhausted hood. For some operations, use cyanide salts only with a hydrogen cyanide gas monitor present and operating.

KEEP AWAY FROM ACIDS. Particular care should be taken to avoid skin contact. If hydrogen cyanide poisoning is suspected, remove the exposed individual from the contaminated area as rapidly as possible and call the emergency number. Special emergency medical antidotes may be needed.

FACESHIELD AND GLASSES   ACID GLOVES   APRON AND ARMGUARDS   AIR EXHAUSTED

Cyanogen

[460–19–5]

CNCN

**POISON/FLAMMABLE/
CORROSIVE (ACID)**

HAZARDS

Gas Eye, nose, and throat irritation occurs at 16 ppm. Bitter almond odor detectable at 250 ppm. Reacts with water to form hydrogen cyanide (see Hydrogen Cyanide). Less toxic than hydrogen cyanide. Flammable.

TLV/PEL 10 ppm

ER 6.6 – 32%

| POISON | FLAMMABLE | CORROSIVE | GAS HAZARD |

PRECAUTIONS

Cyanogen is present in the residues of some plasma aluminum etchers. Etcher residues should be kept within the confines of an air-exhausted hood or used with local exhaust ventilation.

When working with residues from plasma aluminum etchers, use solvent gloves, faceshield and safety glasses, apron and armguards.

Keep away from ignition sources such as flame, spark, and heat.

| FACESHIELD AND GLASSES | SOLVENT GLOVES | APRON AND ARMGUARDS | AIR EXHAUSTED | NO IGNITION SOURCE |

Cyanogen Chloride

CNCl

[506–77–4]

**POISON/
CORROSIVE (ACID)**

HAZARDS

Gas Severe eye and nose irritation occurs at 0.7 ppm. Pungent odor detectable at or below 1 ppm. Prolonged exposures can cause hoarseness, redness of the eyes, and fluid in the eyelids. 2 ppm for 10 minutes is an intolerable concentration. Cyanogen chloride causes toxic effects similar to hydrogen cyanide (see Hydrogen Cyanide). 48 ppm of cyanogen chloride can be fatal in 30 minutes.

TLV/PEL 0.3 ppm C

| POISON | CORROSIVE | GAS HAZARD |

(Manufacturer)

4 / 4 / 2

PRECAUTIONS

Cyanogen chloride is present in the residues of some plasma aluminum etchers. Etcher residues should be kept within the confines of an air-exhausted hood or used with local exhaust ventilation.

When working with residues from plasma aluminum etchers, use solvent gloves, faceshield and safety glasses, apron and armguards.

| FACESHIELD AND GLASSES | SOLVENT GLOVES | APRON AND ARMGUARDS | AIR EXHAUSTED | NO IGNITION SOURCE |

Cyclohexane

[110–82–7]

C_6H_{12}

FLAMMABLE

HAZARDS

Liquid May cause eye irritation. Repeated contact may cause skin irritation.

Vapor Pungent odor detectable at 0.5 to 780 ppm. Concentrations of about 300 ppm can cause some irritation of eyes, nose, and throat. Higher concentrations may cause fatigue, dizziness, nausea, and intoxication. Repeated exposure to high concentrations may cause liver and kidney damage. Flammable.

TLV/PEL	300 ppm	**FP**	30°F
IDLH	1300 ppm	**AT**	473°F
VP	95 mm	**ER**	1.3 – 8.4%

FLAMMABLE

PRECAUTIONS

Avoid skin and eye contact. Use solvent gloves, faceshield and safety glasses, apron and armguards. Use only within the confines of an air-exhausted hood.

Keep away from ignition sources such as flame, spark, and heat.

FACESHIELD AND GLASSES  SOLVENT GLOVES  APRON AND ARMGUARDS  AIR EXHAUSTED  NO IGNITION SOURCE

Cyclohexanone

$C_6H_{10}O$

[108–94–1]

COMBUSTIBLE

HAZARDS

Liquid Can be readily absorbed through the skin. May cause eye irritation. Repeated contact may cause skin irritation.

Vapor Pungent odor detectable at 0.1 to 100 ppm. Concentrations of 50 ppm can cause throat irritation. Higher concentrations may cause fatigue, dizziness, nausea, and intoxication. Repeated exposure to high concentrations may cause liver and kidney damage. Flammable.

TLV/PEL	25 ppm	**FP**	111°F
IDLH	700 ppm	**AT**	788°F
VP	2 mm	**ER**	1.1% – ?

PRECAUTIONS

Avoid skin and eye contact. Use solvent gloves, faceshield and safety glasses, apron and armguards. Use only within the confines of an air-exhausted hood.

Keep away from open flame and high temperatures.

Diacetone Alcohol [123–42–2]

$(CH_3)_2C(OH)CH_2COCH_3$ **FLAMMABLE**
Diacetone

HAZARDS

Liquid Repeated contact can cause skin and eye irritation.

Vapor Mild, sweet odor detectable at 0.3 to 2 ppm. Concentrations of about 100 ppm can cause irritation of eyes, nose, and throat. High concentrations can cause intoxication. Repeated exposure to high concentrations may cause liver damage. Flammable.

TLV/PEL	50 ppm	**FP**	55°F
IDLH	1800 ppm	**AT**	1118°F
VP	1 mm	**ER**	1.8 – 6.9%

FLAMMABLE

PRECAUTIONS

Avoid skin and eye contact. Use solvent gloves, faceshield and safety glasses, apron and armguards. Use only within the confines of an air-exhausted hood.
Keep away from ignition sources such as flame, spark, and heat.

FACESHIELD AND GLASSES   SOLVENT GLOVES   APRON AND ARMGUARDS   AIR EXHAUSTED   NO IGNITION SOURCE

Diborane

[19287–45–7]

B_2H_6
Boroethane
Boron Hydride

POISON/FLAMMABLE

HAZARDS

Gas Burnt chocolate, sickly sweet odor detectable at 1.8 to 3.5 ppm. Can be irritating to eyes, nose, and lungs. Can be readily absorbed through the skin. Can cause headache, light-headedness, nausea, tightness in chest, and fluid in lungs. May cause changes in the central nervous system. May be fatal if absorbed through skin or inhaled. Flammable.

TLV/PEL 0.1 ppm	**FP** –130°F	**ER** 0.8 – 93%
IDLH 15 ppm	**AT** 100 – 125°F	

Important Hazard Toxic at low concentrations. Extremely flammable. Can be toxic at concentrations below the odor threshold. Pyrophoric — ignites spontaneously upon contact with air — above 100°F. Evolves hydrogen and may ignite on contact with water or moisture in air.

POISON   FLAMMABLE   GAS HAZARD   SPECIAL HAZARD   AIR REACTIVE

PRECAUTIONS

Use only within a closed, well-exhausted system. Use only with a toxic gas monitor present and operating.

Keep away from ignition sources such as flame, spark, and heat. All suspected exposures must be reported to the medical department immediately. Keep away from water.

SAFETY GLASSES   AIR EXHAUSTED   NO IGNITION SOURCE   WATER REACTIVE

o-Dichlorobenzene

[95–50–1]

$C_6H_4Cl_2$
1,2-Dichlorobenzene

COMBUSTIBLE

HAZARDS

Liquid Repeated contact can cause skin and eye irritation and possible skin sensitization.

Vapor Mothball-like odor detectable at 0.02 to 4 ppm. Eye irritation noticeable at 25 to 50 ppm and painful to some at 60 to 100 ppm if exposures are for more than a few minutes. Nose and throat irritation, headache, fatigue, and nausea can occur at higher concentrations. Repeated exposure may cause liver and kidney damage. Combustible.

TLV/PEL	25 ppm	**FP**	151°F
IDLH	200 ppm	**AT**	342°F
STEL	50 ppm	**ER**	2.2 – 9.2%
VP	1.2 mm		

COMBUSTIBLE

PRECAUTIONS

Avoid skin and eye contact. Use solvent gloves, faceshield and safety glasses, apron and armguards. Use only within the confines of an air-exhausted hood.

Keep away from open flame and high temperatures.

FACESHIELD AND GLASSES — SOLVENT GLOVES — APRON AND ARMGUARDS — AIR EXHAUSTED — NO IGNITION SOURCE

p-Dichlorobenzene

[106–46–7]

$C_6H_4Cl_2$
1,4-Dichlorobenzene

COMBUSTIBLE

HAZARDS

Liquid Repeated contact can cause skin and eye irritation.

Vapor Mothball-like odor detectable at 0.1 to 15 ppm. Can cause eye, nose, and throat irritation. Repeated exposure may cause liver and kidney damage. Combustible.

TLV/PEL 10 ppm   **VP** 0.64 mm

IDLH 150 ppm   **FP** 150°F

Important Hazard Suspected carcinogen.

COMBUSTIBLE   CANCER CAUSING

2 / 2 / 0

PRECAUTIONS

Avoid skin and eye contact. Use solvent gloves, faceshield and safety glasses, apron and armguards. Use only within the confines of an air-exhausted hood.

Keep away from open flame and high temperatures.

FACESHIELD AND GLASSES   SOLVENT GLOVES   APRON AND ARMGUARDS   AIR EXHAUSTED   NO IGNITION SOURCE

trans 1,2-Dichloroethylene [156–60–5]

$C_2H_2Cl_2$ **FLAMMABLE**

HAZARDS

Liquid Repeated contact can cause skin and eye irritation.

Vapor Pleasant, ether-like odor detectable at 0.08 ppm. Exposure at 2200 ppm can cause burning of the eyes, dizziness, and nausea. Repeated exposure may cause liver damage. Flammable.

TLV/PEL	200 ppm	**AT**	860°F
IDLH	1000 ppm	**VP**	0.64 mm
ER	9.7–12.8%	**FP**	< 36°F

Important Hazard Suspected carcinogen.

COMBUSTIBLE   CANCER CAUSING

PRECAUTIONS

Avoid skin and eye contact. Use solvent gloves, faceshield and safety glasses, apron and armguards. Use only within the confines of an air-exhausted hood.

Keep away from ignition sources such as flame, spark, and heat.

FACESHIELD AND GLASSES   SOLVENT GLOVES   APRON AND ARMGUARDS   AIR EXHAUSTED   NO IGNITION SOURCE

Dichlorosilane

[4109–96–0]

DCS
SiH$_2$Cl$_2$

FLAMMABLE/CORROSIVE

HAZARDS

Gas Can be very irritating and cause chemical burns to eyes and lungs. Fumes in presence of moisture. Reacts with water or moisture in the air to form hydrogen chloride (see Hydrogen Chloride).

FP –62°F

AT 111°F

ER 4.7 – 96%

Important Hazard Extremely flammable.

FLAMMABLE   GAS HAZARD   SPECIAL HAZARD

4 / 4 / 2 / W

PRECAUTIONS

Use only within a closed, well-exhausted system. Keep away from ignition sources such as flame, spark, and heat.

Fire extinguishers are ineffective in controlling fires and gas leaks; the gas supply must be turned off.

Keep away from water including small amounts of moisture.

SAFETY GLASSES   AIR EXHAUSTED   NO IGNITION SOURCE   WATER REACTIVE

Diesel Oil

[77650–28–3]

Fuel Oil No. 2

COMBUSTIBLE

HAZARDS

Liquid Repeated contact can cause skin irritation. Can cause eye irritation.

Vapor Mild petroleum odor. High concentrations can cause headaches, dizziness, and nausea. Higher concentrations can cause vomiting and loss of coordination. Combustible.

FP 100 – 190°F

AT 494°F

ER 0.6 – 7.5%

COMBUSTIBLE

(Manufacturer)

PRECAUTIONS

Avoid skin and eye contact. Use solvent gloves, faceshield and safety glasses, apron and armguards. Use only within the confines of an air-exhausted hood.

Keep away from open flame and high temperatures.

| FACESHIELD AND GLASSES | SOLVENT GLOVES | APRON AND ARMGUARDS | AIR EXHAUSTED | NO IGNITION SOURCE |

Diethylene Glycol

[111–46–6]

$O(CH_2CH_2OH)_2$
DEG

COMBUSTIBLE

HAZARDS

Mist/Vapor High concentrations may cause headache, dizziness, and nausea. Repeated exposure may cause liver and kidney damage. Combustible.

- **VP** < 0.01 mm
- **ER** 3 – 7%
- **FP** 255°F
- **AT** 444°F

COMBUSTIBLE

PRECAUTIONS

Particular care should be taken to avoid skin and eye contact. Use solvent gloves, faceshield and safety glasses, apron and armguards. Use in a well-ventilated area.

Keep away from ignition sources such as flame, spark, and heat.

SOLVENT GLOVES

SAFETY GLASSES

NO IGNITION SOURCE

Diethylene Glycol Dimethyl Ether*

[111–96–6]

COMBUSTIBLE

CH$_3$(OCH$_2$CH$_2$)$_2$OCH$_3$
Diglyme
Bis(2-Methoxyethyl)Ether

HAZARDS

Liquid May cause eye irritation. Repeated contact may cause skin irritation.

Vapor Mild odor. High concentrations may cause dizziness, drowsiness, and headache. Combustible.

VP approx. 3 mm   **ER** 1.5 – 17.4%

FP 153°F

Important Hazard Suspected of causing reproductive disorders.

COMBUSTIBLE   SPECIAL HAZARD

PRECAUTIONS

Avoid skin and eye contact. Use solvent gloves, faceshield and safety glasses, apron and armguards. Use only within the confines of an air-exhausted hood.
Keep away from open flame and high temperatures.

FACESHIELD AND GLASSES   SOLVENT GLOVES   APRON AND ARMGUARDS   AIR EXHAUSTED   NO IGNITION SOURCE

* The Semiconductor Industry Association recommends that its member companies eliminate the use of this chemical from their manufacturing operations.

Diethylene Glycol Monobutyl Ether

[112–34–5]

$C_8H_{17}O_2OH$
Butyl Carbitol

COMBUSTIBLE

HAZARDS

Liquid Can be readily absorbed through the skin. Repeated contact can cause skin irritation. Can cause severe eye irritation.

Vapor/Mist If the liquid is heated, vapors may be irritating to eyes, nose, and throat. Exposure to high concentrations can cause dizziness. This chemical is not likely to form significant concentrations of vapors at room temperature. Combustible.

VP 0.02 mm

FP 172°F

AT 400°F

ER 0.9 – 6.2%

COMBUSTIBLE

2 1 0

PRECAUTIONS

Avoid skin and eye contact. Use solvent gloves, faceshield and safety glasses, apron and armguards. Use only within the confines of an air-exhausted hood.

Keep away from open flame and high temperatures.

FACESHIELD AND GLASSES  | SOLVENT GLOVES  | APRON AND ARMGUARDS 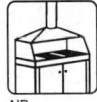 | AIR EXHAUSTED  | NO IGNITION SOURCE

Diethylene Glycol Monoethyl Ether

[111–90–0]

$C_2H_5(OCH_2CH_2)_2OH$
Carbitol

COMBUSTIBLE

HAZARDS

Liquid Can be readily absorbed through the skin. Repeated contact can cause skin irritation.

Vapor If the liquid is heated vapors may be irritating to eyes, nose, and throat. Exposure to high concentrations can cause dizziness. Repeated contact at high concentrations may cause liver and lung damage. This chemical is not likely to form significant concentrations of vapors at room temperature. Combustible.

VP 0.13 mm at 25°C **AT** 400°F
FP 201°F **ER** 1.2–9%

COMBUSTIBLE

PRECAUTIONS

Particular care should be taken to avoid skin and eye contact. Use solvent gloves, faceshield and safety glasses, apron and armguards. Use only within the confines of an air-exhausted hood.

Keep away from open flame and high temperatures.

FACESHIELD AND GLASSES  | SOLVENT GLOVES  | APRON AND ARMGUARDS 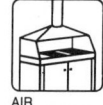 | AIR EXHAUSTED  | NO IGNITION SOURCE

Diethylene Glycol Monoethyl Ether Acetate [112–15–2]

Glycol Ether DE Acetate
Carbitol Acetate

COMBUSTIBLE

HAZARDS

Liquid Repeated contact can cause skin irritation. Can cause severe eye irritation.

Vapor/Mist If the liquid is heated vapors may be irritating to eyes, nose and throat. Exposure to high concentrations can cause dizziness. Repeated contact at high concentrations may cause liver and lung damage. This chemical is not likely to form significant concentrations of vapors at room temperature. Combustible.

VP 0.05 mm **AT** 680°F

FP 211°F **ER** 0.98 – 19.4%

COMBUSTIBLE

1 / 1 / 0

PRECAUTIONS

Avoid skin and eye contact. Use solvent gloves, faceshield and safety glasses, apron and armguards. Use only within the confines of an air-exhausted hood.
Keep away from open flame and high temperatures.

FACESHIELD AND GLASSES | SOLVENT GLOVES | APRON AND ARMGUARDS | AIR EXHAUSTED | NO IGNITION SOURCE

Diethyl Telluride

[627–54–3]

$(C_2H_5)_2Te$

FLAMMABLE

HAZARDS

Liquid Ignites spontaneously in air to form tellurium oxides.

Mist Can cause sleepiness, loss of appetite, nausea, metallic taste, and a garlic odor to the breath and perspiration.

TLV/PEL 0.1 mg/m^3

Important Hazard Pyrophoric — ignites and burns on contact with air. Contact with water or moisture in air may cause fire by release of flammable vapors and heat.

FLAMMABLE	SPECIAL HAZARD	AIR REACTIVE

NFPA: 1 / 4 / 3 / W

PRECAUTIONS

Avoid inhalation and skin and eye contact. Use solvent gloves and safety glasses.

Use only within a closed, well-exhausted system.

Keep away from ignition sources such as flame, spark, and heat.

Keep away from water.

SAFETY GLASSES	SOLVENT GLOVES	AIR EXHAUSTED	NO IGNITION SOURCE	WATER REACTIVE

Dimethyl Acetamide

[127–19–5]

$CH_3CON(CH_3)_2$

COMBUSTIBLE

HAZARDS

Liquid Can be readily absorbed through the skin. Can cause skin and eye irritation.

Vapor Burnt, oily odor detectable at 21 to 47 ppm. Can cause liver damage. Combustible.

TLV/PEL	10 ppm	**FP**	158°F
IDLH	300 ppm	**AT**	914°F
VP	1.5 mm	**ER**	1.8 – 11.5%

Important Hazard Odor or irritation does not provide an indication of potentially harmful concentrations. May cause reproductive disorders.

COMBUSTIBLE   SPECIAL HAZARD

PRECAUTIONS

Particular care should be taken to avoid skin and eye contact. Use solvent gloves, faceshield and safety glasses, apron and armguards. Use only within the confines of an air-exhausted hood.

Keep away from open flame and high temperatures.

FACESHIELD AND GLASSES   SOLVENT GLOVES   APRON AND ARMGUARDS   AIR EXHAUSTED   NO IGNITION SOURCE

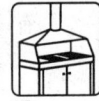

Dimethyl Formamide

[68–12–2]

HCON(CH$_3$)$_2$

COMBUSTIBLE

HAZARDS

Liquid Can be readily absorbed through the skin. Can cause skin and eye irritation.

Vapor Fishy odor detectable at 0.05 to 0.3 ppm. Can be irritating to eyes and lungs. Can cause nausea and vomiting. Repeated contact at high concentrations can cause liver damage. Combustible.

TLV/PEL	10 ppm	**FP**	136°F
IDLH	500 ppm	**AT**	833°F
VP	2.7 mm	**ER**	2.2 – 15.2%

Important Hazard Suspected of causing reproductive disorders.

COMBUSTIBLE SPECIAL HAZARD

PRECAUTIONS

Particular care should be taken to avoid skin and eye contact. Use solvent gloves, faceshield and safety glasses, apron and armguards. Use only within the confines of an air-exhausted hood.

Keep away from open flame and high temperatures.

FACESHIELD AND GLASSES   SOLVENT GLOVES   APRON AND ARMGUARDS   AIR EXHAUSTED   NO IGNITION SOURCE

Dimethyl Sulfoxide

[67–68–5]

COMBUSTIBLE

$(CH_3)_2SO$
DMSO
Methyl Sulfoxide

HAZARDS

Liquid Can be readily absorbed through the skin. Can cause skin and eye irritation. May cause eye damage.

Vapor Combustible.

VP 0.37 mm

FP 203°F

AT 419°F

ER 2.6 – 29%

Important Hazard Suspected of causing reproductive disorders. May cause allergies in certain sensitive individuals.

COMBUSTIBLE SPECIAL HAZARD

PRECAUTIONS

Particular care should be taken to avoid skin and eye contact. Use solvent gloves, faceshield and safety glasses, apron and armguards. Use only within the confines of an air-exhausted hood.
Keep away from open flame and high temperatures.

FACESHIELD AND GLASSES   SOLVENT GLOVES   APRON AND ARMGUARDS   AIR EXHAUSTED   NO IGNITION SOURCE

Dioxane

[123–91–1]

FLAMMABLE

$C_4H_8O_2$
1-4-Diethylene Dioxide

HAZARDS

Liquid Can be readily absorbed through the skin. Repeated contact can cause skin and eye irritation.

Vapor Mild, ether-like odor detectable at 0.8 to 6 ppm. At concentrations of about 300 ppm, can cause irritation of eyes, nose, and throat. Repeated exposure at high concentrations can cause liver and kidney damage. Flammable.

TLV/PEL	25 ppm	**FP**	54°F
IDLH	500 ppm	**AT**	356°F
VP	29 mm	**ER**	2 – 22.2%

Important Hazard Suspected carcinogen.

FLAMMABLE 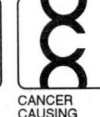 CANCER CAUSING

PRECAUTIONS

Particular care should be taken to avoid skin and eye contact. Use solvent gloves, faceshield and safety glasses, apron and armguards. Use only within the confines of an air-exhausted hood.

Keep away from ignition sources such as flame, spark, and heat.

FACESHIELD AND GLASSES    SOLVENT GLOVES    APRON AND ARMGUARDS    AIR EXHAUSTED    NO IGNITION SOURCE

Dipropylene Glycol Methyl Ether

[34590–94–8]

COMBUSTIBLE

$CH_3OC_3H_6OC_3H_6OH$
DPGME

HAZARDS

Liquid Can be readily absorbed through the skin. May cause skin and eye irritation.

Vapor Mild, sweet odor detectable at 35 ppm. Exposure to concentrations above 100 ppm can be irritating to eyes, nose, and throat and would not be tolerated willingly. Exposure to high concentrations can cause dizziness. Combustible.

TLV/PEL	100 ppm	**VP**	0.3 mm
IDLH	600 ppm	**FP**	185°F
STEL	150 ppm	**ER**	1.1 – 3.0%

COMBUSTIBLE

PRECAUTIONS

Avoid skin and eye contact. Use solvent gloves, faceshield and safety glasses, apron and armguards. Use only within the confines of an air-exhausted hood.

Keep away from open flame and high temperatures.

FACESHIELD AND GLASSES SOLVENT GLOVES APRON AND ARMGUARDS AIR EXHAUSTED NO IGNITION SOURCE

Dross, Solder

Inorganic Lead and Tin **POISON**

HAZARDS

Dust Inhalation and ingestion of lead compounds can damage blood cells and nervous system. Can cause headache, insomnia, dizziness, and anemia. Can cause skin and eye irritation.

TLV/PEL 0.05 mg/m^3 (as lead)

IDLH 700 mg/m^3 (as lead)

Important Hazard Cumulative poison. Can cause reproductive disorders. Suspected carcinogen. Emits toxic lead fumes when heated above 800°F.

| POISON |  SPECIAL HAZARD |  CANCER CAUSING | (as Lead, Manufacturer) |

PRECAUTIONS

Avoid inhalation and skin and eye contact. Use acid or solvent gloves and safety glasses. Use only within the confines of an air-exhausted hood. Brushing, sweeping, heating, or scraping of dross must be done in an air-exhausted hood or with local exhaust.

Wash hands thoroughly after handling and before eating, drinking, or smoking.

FACESHIELD AND GLASSES  · ACID GLOVES  · APRON AND ARMGUARDS 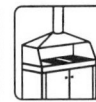 · AIR EXHAUSTED

Epoxy Resins

VARIES WITH THE TYPE OF EPOXY

HAZARDS

Liquid   May cause skin and eye irritation. Repeated contact may cause skin rash and an allergic skin reaction.

Vapor   May be irritating to skin, eyes, and lungs.

Important Hazard   Contact the safety department for hazard information on the particular resin you are working with.

SPECIAL HAZARD

PRECAUTIONS

Particular care should be taken to avoid skin and eye contact. Use solvent gloves and safety glasses. Use in a well-ventilated area. Some resins may require an air-exhausted hood.

SAFETY GLASSES   SOLVENT GLOVES   AIR EXHAUSTED

Ethanolamine

[141–43–5]

HOCH$_2$CH$_2$NH$_2$
Monoethanolamine
MEA

CORROSIVE (BASE)/ COMBUSTIBLE

HAZARDS

Liquid Can cause chemical burns to skin and eyes.

Vapor Ammonia-like odor detectable at 3 to 4 ppm. Can cause skin, eye, and lung irritation. High concentrations may cause severe lung damage. Repeated contact at high concentrations may cause liver and kidney damage. Combustible.

TLV/PEL	3 ppm	**FP**	185°F
STEL	6 ppm	**AT**	770°F
IDLH	30 ppm	**ER**	5.5 – 17%
VP	4 mm		

CORROSIVE COMBUSTIBLE

3 / 2 / 0

PRECAUTIONS

Avoid skin and eye contact. Use solvent gloves, faceshield and safety glasses, apron and armguards. Use only within the confines of an air-exhausted hood.

Keep away from open flame and high temperatures.

FACESHIELD AND GLASSES   SOLVENT GLOVES   APRON AND ARMGUARDS   AIR EXHAUSTED   NO IGNITION SOURCE

2-Ethoxyethanol*

[110–80–5]

$C_2H_5OCH_2CH_2OH$
Ethylene Glycol Monoethyl Ether

COMBUSTIBLE

HAZARDS

Liquid Can be readily absorbed through the skin. Can cause skin and eye irritation.

Vapor Sweet odor detectable at 0.3 to 1.3 ppm. Can damage blood cells. May cause damage to lungs, kidneys, liver, and spleen. Combustible.

TLV/PEL	5 ppm	**FP**	112°F
IDLH	500 ppm	**AT**	455°F
VP	3.7 mm	**ER**	1.7 – 16%

Important Hazard Suspected of causing reproductive disorders.

COMBUSTIBLE   SPECIAL HAZARD

2 / 2 / 0

PRECAUTIONS

Particular care should be taken to avoid skin and eye contact. Use solvent gloves, faceshield and safety glasses, apron and armguards. Use only within the confines of an air-exhausted hood.
Keep away from open flame and high temperatures.

FACESHIELD AND GLASSES   SOLVENT GLOVES   APRON AND ARMGUARDS   AIR EXHAUSTED   NO IGNITION SOURCE

* The Semiconductor Industry Association recommends that its member companies eliminate the use of this chemical in their manufacturing operations.

2-Ethoxyethyl Acetate*

[111–15–9]

COMBUSTIBLE

Ethylene Glycol Monoethyl Ether
  Acetate
EGMEA

HAZARDS

Liquid Can be readily absorbed through the skin. Repeated contact can cause eye and skin irritation.

Vapor Sweet odor detectable at 0.06 to 0.3 ppm. High concentrations can cause irritation of eyes, nose, and throat and dizziness. Prolonged exposures may cause kidney damage. Combustible.

TLV/PEL 5 ppm	**VP** 1.2 mm	**AT** 715°F
IDLH 500 ppm	**FP** 117°F	**ER** 1.7 – ?%

Important Hazard Suspected of causing reproductive disorders.

COMBUSTIBLE   SPECIAL HAZARD

2 / 2 / 0

PRECAUTIONS

Particular care should be taken to avoid skin and eye contact. Use solvent gloves, faceshield and safety glasses, apron and armguards. Use only within the confines of an air-exhausted hood.
Keep away from open flame and high temperatures.

FACESHIELD AND GLASSES   SOLVENT GLOVES   APRON AND ARMGUARDS   AIR EXHAUSTED   NO IGNITION SOURCE

* The Semiconductor Industry Association recommends that its member companies eliminate the use of this chemical in their manufacturing operations.

Ethyl Acetate

[141–78–6]

CH$_3$COOC$_2$H$_5$

FLAMMABLE

HAZARDS

Liquid Repeated contact can cause skin and eye irritation.

Vapor Fruity odor detectable at 6 to 13 ppm. Exposures at about 400 ppm can cause irritation to eyes, nose, gums, and throat. Inhalation of high concentrations may cause headache, dizziness, and drowsiness. Flammable.

TLV/PEL	400 ppm
IDLH	2000 ppm
VP	76 mm
FP	24°F
AT	800°F
ER	2.0 – 12%

FLAMMABLE

PRECAUTIONS

Avoid skin and eye contact. Use solvent gloves, faceshield and safety glasses, apron and armguards. Use only within the confines of an air-exhausted hood.

Keep away from ignition sources such as flame, spark, and heat.

FACESHIELD AND GLASSES | SOLVENT GLOVES | APRON AND ARMGUARDS | AIR EXHAUSTED | NO IGNITION SOURCE

Ethyl Alcohol

[64–17–5]

C_2H_5OH
Ethanol

FLAMMABLE

HAZARDS

Liquid Repeated contact can cause skin and eye irritation.

Vapor Fragrant odor detectable at 5 to 10 ppm. Can be irritating to eyes and lungs. Inhalation of very high concentrations can cause headache and dizziness. Flammable.

TLV/PEL	1000 ppm	**AT**	685°F
IDLH	3300 ppm	**VP**	44 mm
ER	3.3 – 19%	**FP**	55°F

Important Hazard Suspected of causing reproductive disorders (based on studies in which alcohol is consumed in large quantities).

FLAMMABLE   SPECIAL HAZARD

PRECAUTIONS

Avoid skin and eye contact. Use solvent gloves, faceshield and safety glasses, apron and armguards. Use only within the confines of an air-exhausted hood.

Keep away from ignition sources such as flame, spark, and heat.

FACESHIELD AND GLASSES   SOLVENT GLOVES   APRON AND ARMGUARDS   AIR EXHAUSTED   NO IGNITION SOURCE

215

Ethyl Benzene

[100–41–4]

$C_6H_5(C_2H_5)$
Phenylethane

FLAMMABLE

HAZARDS

Liquid Repeated contact can cause skin and eye irritation.

Vapor Distinctive aromatic ("airplane glue") odor detectable at 2 to 20 ppm. Exposure at 100 ppm can cause mild vertigo, sleepiness, and headache; at 200 ppm can cause mild eye irritation; at 1000 to 2000 ppm for 6 minutes can cause fatigue and increasing vertigo, chest constriction, dizziness, and eye irritation and tearing. Repeated exposure to high concentrations may cause kidney and liver damage. Flammable.

TLV/PEL	100 ppm	**FP**	59°F
STEL	125 ppm	**AT**	810°F
IDLH	800 ppm	**ER**	1 – 6.8%
VP	7.1 mm		

FLAMMABLE

PRECAUTIONS

Avoid skin and eye contact. Use solvent gloves, faceshield and safety glasses, aprons and armguards. Use only within the confines of an air-exhausted hood.

Keep away from ignition sources such as flame, spark, and heat.

FACESHIELD AND GLASSES · SOLVENT GLOVES · APRON AND ARMGUARDS · AIR EXHAUSTED · NO IGNITION SOURCE

Ethylenediamine

[107–15–3]

NH$_2$CH$_2$CH$_2$NH$_2$
1,2-Diaminoethane

**CORROSIVE (BASE)/
FLAMMABLE**

HAZARDS

Liquid Can cause burns to skin and eyes. Can be readily absorbed through the skin.

Vapor Ammonia-like odor detectable at 1 to 11 ppm. Can cause irritation of eyes, nose, and throat. Exposure at 200 ppm can cause slight nose irritation and tingling of face. High concentrations may cause lung, kidney, and liver damage. Combustible.

TLV/PEL	10 ppm	**FP**	91°F
IDLH	1000 ppm	**AT**	725°F
VP	10 mm	**ER**	4.2 – 14.4%

Important Hazard Can cause allergic respiratory reaction or allergic skin sensitization in some susceptible individuals.

CORROSIVE   FLAMMABLE   SPECIAL HAZARD

3 / 3 / 0

PRECAUTIONS

Avoid skin and eye contact.
Use acid gloves, faceshield and safety glasses, apron and armguards. Use only within the confines of an air-exhausted hood.
Keep away from open flame and high temperatures.

FACESHIELD AND GLASSES   ACID GLOVES   APRON AND ARMGUARDS   AIR EXHAUSTED   NO IGNITION SOURCE

Ethylenediaminetetraacetic Acid

[60-00-4]

EDTA
Ethylene Dinitrilotetraacetic Acid

UNCLASSIFIED

HAZARDS

Dust/Mist May cause irritation of skin, eyes, nose, throat, and lungs.

(Manufacturer)

PRECAUTIONS

Avoid skin and eye contact. Use solvent gloves and safety glasses. Use in a well-ventilated area.

SAFETY GLASSES

SOLVENT GLOVES

Ethylene Glycol

[107–21–1]

COMBUSTIBLE

CH_2OHCH_2OH
1,2-Ethanediol
Ethylene Alcohol

HAZARDS

Liquid Repeated contact can cause skin and eye irritation.

Vapor Odor detectable at 0.08 to 25 ppm. This chemical is not likely to form significant concentrations of vapors at room temperature (66 ppm is the maximum concentration at 20°C). Exposure to 56 ppm can not be tolerated for long because of throat irritation. Combustible.

TLV/PEL	50 ppm C
VP	0.05 mm
FP	232°F
AT	748°F
ER	3.2 – 15.3%

COMBUSTIBLE

PRECAUTIONS

Avoid skin and eye contact. Use solvent gloves, faceshield and safety glasses, apron and armguards. Use in a well-ventilated area.
Keep away from open flame and high temperatures.

FACESHIELD AND GLASSES   SOLVENT GLOVES   APRON AND ARMGUARDS   NO IGNITION SOURCE

219

Ethyl Ether

[60–29–7]

$C_2H_5OC_2H_5$
Diethylether
Ether

FLAMMABLE

HAZARDS

Liquid Repeated contact can cause skin and eye irritation.

Vapor Sweetish odor detectable at 0.1 – 7 ppm. 200 ppm can cause nose irritation. Concentrations at and above 2000 ppm for extended periods of time may cause dizziness and vomiting. High concentrations may also cause eye and throat irritation, headache, and either depression or excitation. Repeated exposure may cause loss of appetite, nausea, shortness of breath, irritability, and blood and heart abnormalities. Flammable.

TLV/PEL	400 ppm	**VP**	442 mm	**ER**	1.7 – 48%
STEL	500 ppm	**FP**	–49°F		
IDLH	1900 ppm	**AT**	320°F		

Important Hazard Vapors are extremely flammable.

FLAMMABLE SPECIAL HAZARD

PRECAUTIONS

Avoid skin and eye contact. Use solvent gloves, faceshield and safety glasses, aprons and armguards. Use only within the confines of an air-exhausted hood.
Keep away from ignition sources such as flame, spark, and heat.

FACESHIELD AND GLASSES SOLVENT GLOVES APRON AND ARMGUARDS AIR EXHAUSTED NO IGNITION SOURCE

Ethyl Lactate

[97–64–3]

$CH_3CHOHCOOC_2H_5$ **COMBUSTIBLE**

HAZARDS

Liquid May cause skin and eye irritation.

Vapor Mild odor. May be irritating to eyes and lungs. Combustible.

FP 115°F

AT 752°F

COMBUSTIBLE

PRECAUTIONS

Avoid skin and eye contact. Use solvent gloves, faceshield and safety glasses, apron and armguards. Use only within the confines of an air-exhausted hood.

Keep away from open flame and high temperatures.

FACESHIELD AND GLASSES SOLVENT GLOVES APRON AND ARMGUARDS AIR EXHAUSTED

Ferric Chloride

[7705–08–0]

FeCl$_3$ **UNCLASSIFIED**

HAZARDS

Solid/Liquid   Can cause skin and eye irritation.

Dust/Mist   Can cause severe irritation of the eyes, nose, throat, and lungs. High concentrations may cause fluid in lungs.

TLV/PEL   1 mg/m^3 (as iron)

(Solid, Manufacturer)   (Solution, Manufacturer)

PRECAUTIONS

Avoid skin and eye contact. Use acid or solvent gloves and safety glasses.
Use in a well-ventilated area.
Keep solid away from water.

ACID GLOVES   SAFETY GLASSES   WATER REACTIVE

Ferrous Chloride

[7758–94–3]

FeCl$_2$

CORROSIVE

HAZARDS

Solid/Liquid   Can cause skin and eye irritation.

Dust/Mist   Can cause irritation of the nose, throat, and lungs.

TLV/PEL   1 mg/m^3 (as iron)

(Manufacturer)

PRECAUTIONS

Avoid skin and eye contact. Use acid or solvent gloves and safety glasses. Use in a well-ventilated area.

ACID GLOVES

SAFETY GLASSES

Ferrous Sulfate

[7720–78–7]

FeSO$_4$

UNCLASSIFIED

HAZARDS

Solid/Liquid May cause skin and eye irritation.

Dust/Mist May cause irritation of the nose, throat, and lungs.

TLV/PEL 1 mg/m^3 (as iron)

(Manufacturer)

PRECAUTIONS

Avoid skin and eye contact. Use acid or solvent gloves and safety glasses. Use in a well-ventilated area.

ACID GLOVES   SAFETY GLASSES

Fluoboric Acid, 48%

[16872–11–0]

CORROSIVE (ACID)

HBF$_4$
Fluoroboric Acid
Hydrogen Tetrafluoroborate

HAZARDS

Liquid Can cause severe burns to skin and eyes.

Vapor Pungent odor. Can cause irritation of the nose, throat, and lungs. High concentrations can cause fluid in lungs. Repeated exposure to inorganic fluoride compounds may cause bone damage.

TLV/PEL 2.5 mg/m^3 (as F)

VP 5 – 10 mm

CORROSIVE

3 / 0 / 1 (Manufacturer)

PRECAUTIONS

Avoid skin and eye contact. Use acid gloves, faceshield and safety glasses, apron and armguards. Use only within the confines of an air-exhausted hood.

FACESHIELD AND GLASSES   ACID GLOVES   APRON AND ARMGUARDS   AIR EXHAUSTED

Fluorine

[7782–41–4]

F_2

POISON

HAZARDS

Gas Pale yellow to greenish gas with a sharp odor detectable at 0.1 to 0.2 ppm. Exposures of 10 ppm for 15 minutes cause minimal irritation. Brief exposures to 25 ppm can be tolerated, but can cause sore throat and chest pain which persists for several hours. 50 ppm is intolerable. High concentrations can cause constriction of the throat and lungs followed by the delayed onset of fluid in the lungs.

TLV/PEL 1 ppm   **IDLH** 25 ppm

STEL 2 ppm

Important Hazard Poisonous if inhaled. Can be fatal at high concentrations. Effects of exposure may not be immediately evident. Will react with water to form hydrogen fluoride and oxygen.

POISON   GAS HAZARD   OXIDIZER   SPECIAL HAZARD

PRECAUTIONS

Use only within a closed, air-exhausted system. Avoid all direct contact. Keep away from other chemicals particularly water.

SAFETY GLASSES   AIR EXHAUSTED   WATER REACTIVE

Formaldehyde [50–00–0]

HCHO **COMBUSTIBLE**

HAZARDS

Liquid Can be irritating to skin and eyes. May cause an allergic skin rash.

Gas/Vapor Strong odor detectable at 0.04 to 1 ppm. Exposure at 0.2 ppm may cause mild eye irritation. Tingling of eyes, nose, and throat occurs in most people at 2 to 3 ppm. Most people can tolerate 4 to 5 ppm for 10 to 30 minutes. At 10 to 20 ppm breathing is difficult and cough and intolerable tearing occurs. Serious injury is likely to occur at 50 to 100 ppm. May cause cancer of the nose. Combustible.

TLV/PEL	0.3 ppm C		**FP**	133 – 185°F
STEL	2 ppm		**AT**	795°F
IDLH	20 ppm		**ER**	7 – 73%

Important Hazard Suspected carcinogen.

COMBUSTIBLE    GAS HAZARD 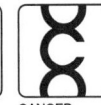   CANCER CAUSING

(Formaldehyde Solutions) 3-2-0   (Formaldehyde) 3-4-0

PRECAUTIONS

Avoid skin and eye contact. Use solvent gloves, faceshield and safety glasses, apron and armguards. Use only within the confines of an air-exhausted hood.

Keep away from open flame and high temperatures.

FACESHIELD AND GLASSES    SOLVENT GLOVES    APRON AND ARMGUARDS 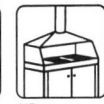   AIR EXHAUSTED    NO IGNITION SOURCE

Formic Acid

[64–18–6]

HCOOH
Methanoic Acid

**CORROSIVE (ACID)/
COMBUSTIBLE**

HAZARDS

Liquid Can cause chemical burns to skin and eyes.

Vapor Strong, penetrating odor detectable at 10 to 20 ppm. Exposure at 15 ppm can cause severe irritation of nose and throat, runny nose, coughing, and difficulty breathing. Symptoms of fluid in the lungs, such as shortness of breath, may not occur until a few hours after exposure. Combustible.

TLV/PEL	5 ppm	**FP**	122°F
STEL	10 ppm	**AT**	813°F
IDLH	30 ppm	**ER**	18 – 57%
VP	35 mm		

CORROSIVE    COMBUSTIBLE

NFPA: 3 / 2 / 0

PRECAUTIONS

Avoid skin and eye contact. Use acid gloves, faceshield and safety glasses, apron and armguards. Use only within the confines of an air-exhausted hood. Keep away from open flame and high temperatures.

FACESHIELD AND GLASSES    ACID GLOVES    APRON AND ARMGUARDS    AIR EXHAUSTED    NO IGNITION SOURCE

Freons

UNCLASSIFIED

HAZARDS

Liquid Repeated contact can be irritating to skin and eyes.

Vapor/Gas High concentrations can cause dizziness and lung damage.

TLV/PEL 500 – 1000 ppm

Important Hazard Can emit highly toxic vapors when heated to decomposition. Freon TMC contains 50% methylene chloride (see Methylene Chloride). Freon TMS contains 6% methyl alcohol (see Methyl Alcohol).

NOTE: The use of these chemicals is being phased out because they damage the ozone layer.

SPECIAL HAZARD

PRECAUTIONS

Avoid skin and eye contact. Use solvent gloves and safety glasses. Use in a well-ventilated area.

Avoid high temperatures.

SAFETY GLASSES

SOLVENT GLOVES

Furfuryl Alcohol

[98–00–0]

$C_4H_3OCH_3OH$

COMBUSTIBLE

HAZARDS

Liquid Can be readily absorbed through the skin. Can cause irritation of skin and eyes.

Vapor Faint burning odor detectable at 8 ppm. Exposure at 16 ppm may cause eye irritation and tearing. High concentrations may cause lung congestion, nausea, reduced body temperature, diarrhea, and dizziness. Combustible.

TLV/PEL	10 ppm	**FP**	167°F
STEL	15 ppm	**AT**	915°F
IDLH	75 ppm	**ER**	1.8 – 16.3%
VP	0.4 mm		

COMBUSTIBLE

PRECAUTIONS

Particular care should be taken to avoid skin and eye contact. Use solvent gloves, faceshield and safety glasses, apron and armguards. Use only within the confines of an air-exhausted hood.

Keep away from open flame and high temperatures.

FACESHIELD AND GLASSES — SOLVENT GLOVES — APRON AND ARMGUARDS — AIR EXHAUSTED — NO IGNITION SOURCE

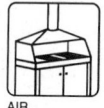

Gallium

Ga

[7440-55-3]

UNCLASSIFIED

HAZARDS

Liquid May cause skin irritation.

Mist Repeated exposure may cause skin irritation and damage bone marrow.

PRECAUTIONS

Avoid skin and eye contact. Use acid or solvent gloves and safety glasses. Use in a well-ventilated area.

SAFETY GLASSES   ACID GLOVES

Gallium Arsenide

[1303–00–0]

GaAs

POISON

HAZARDS

Solid Can be irritating to skin and eyes. Can cause discoloration of skin.

Dust Can be irritating to skin, eyes, and lungs. May cause nose and liver damage. May cause lung and nervous system damage. May cause skin and lung cancer.

TLV/PEL 0.01 mg/m^3 (as arsenic)

Important Hazard Suspected carcinogen. Reacts with steam and acids to form a highly toxic gas, arsine.

POISON   CANCER CAUSING   SPECIAL HAZARD

PRECAUTIONS

Use within a closed, well-exhausted system or within the confines of an air-exhausted hood. Use acid or solvent gloves and safety glasses.

Wash hands thoroughly after handling and before eating, drinking, or smoking.

Keep away from water.

ACID GLOVES   SAFETY GLASSES   AIR EXHAUSTED   WATER REACTIVE

Gasoline

[8006–61–9]

C_4–C_{12} Hydrocarbons
Petroleum Distillate

FLAMMABLE

HAZARDS

Liquid Repeated contact can cause skin and eye irritation and central nervous system depression.

Vapor Odor detectable at 0.06 to 0.15 ppm. 500 to 900 ppm causes eye, nose, and throat irritation and dizziness in 1 hour. Flammable.

TLV/PEL	300 ppm	**FP**	–50°F
STEL	500 ppm	**AT**	495°F
VP	40 mm	**ER**	0.6 – 8.0%

Important Hazard Vapors are extremely flammable. Gasoline can contain up to 5% benzene — a confirmed human carcinogen.

FLAMMABLE   CANCER CAUSING   SPECIAL HAZARD   3 / 1 / 0

PRECAUTIONS

Avoid skin and eye contact. Use solvent gloves, faceshield and safety glasses, apron and armguards. Use only within the confines of an air-exhausted hood.

Keep away from ignition sources such as flame, spark, and heat.

FACESHIELD AND GLASSES   SOLVENT GLOVES   APRON AND ARMGUARDS   AIR EXHAUSTED   NO IGNITION SOURCE

Germanium

[7440–56–4]

Ge

UNCLASSIFIED

HAZARDS

Dust/Mist Repeated exposure may cause skin irritation. Dust can ignite in air.

PRECAUTIONS

Avoid skin and eye contact. Use acid or solvent gloves and safety glasses. Use in a well-ventilated area. Keep away from dusts.

SAFETY GLASSES   ACID GLOVES   NO IGNITION SOURCE

Germanium Tetrahydride [7782–65–2]

GeH$_4$
Germane
Germanium Hydride

POISON/FLAMMABLE

HAZARDS

Gas May damage blood cells. May cause nosebleed, dizziness, headache, nausea/vomiting, liver and kidney damage. Flammable.

TLV/PEL 0.2 ppm

Important Hazard Poisonous at low concentrations. May be toxic at concentrations below the odor threshold. Pyrophoric — may ignite and burn on contact with air. May generate hydrogen and ignite on contact with water.

POISON FLAMMABLE GAS HAZARD SPECIAL HAZARD

4 / 4 / 3 / W

PRECAUTIONS

Use only within a closed, well-exhausted system. Use only with a toxic gas monitor present and operating. All suspected exposures should be reported to the medical department immediately.

Keep away from ignition sources such as flame, spark, and heat. Fire extinguishers are ineffective in controlling fires from gas leaks; the gas supply must be turned off.

SAFETY GLASSES   AIR EXHAUSTED   NO IGNITION SOURCE

Glutamic Acid Hydrochloride

[138–15–8]

$C_5H_9O_4N \cdot HCl$

UNCLASSIFIED

HAZARDS

Dust Can cause eye and skin irritation. Can irritate the nose and throat. When mixed with water it liberates hydrochloric acid (see Hydrogen Chloride).

PRECAUTIONS

Avoid skin and eye contact. Use acid or solvent gloves and safety glasses. Use in a well-ventilated area.

ACID GLOVES   SAFETY GLASSES

Glutaric Acid

[110–94–1]

$C_5H_8O_4$

UNCLASSIFIED

HAZARDS

Dust/Mist May cause eye and skin irritation. High exposures may cause kidney damage.

PRECAUTIONS

Avoid skin and eye contact. Use acid or solvent gloves and safety glasses. Use in a well-ventilated area.

ACID GLOVES SAFETY GLASSES

Glycerin

[56–81–5]

$C_3H_8O_3$
Glycerol

COMBUSTIBLE

HAZARDS

Mist High concentrations may cause nose and throat irritation. Very high concentrations may cause kidney damage.

TLV/PEL	10 mg/m^3
VP	0.0025 mm at 50°C
FP	350°F
AT	698°F

COMBUSTIBLE

1 / 1 / 0

PRECAUTIONS

Avoid skin and eye contact. Use solvent gloves and safety glasses. Use in a well-ventilated area.
Keep away from open flame and high temperatures.

SOLVENT GLOVES   SAFETY GLASSES   NO IGNITION SOURCE

Halons
NONFLAMMABLE GAS

HAZARDS

Gas Acts as a simple asphyxiant by displacing oxygen in the air. Extremely high concentrations can cause headache and dizziness.

TLV/PEL 1000 ppm (as halon 122— dichlorodifluoromethane)

(as Halon 1301 — Bromotrifluoromethane; Manufacturer)

PRECAUTIONS

Use in a well-ventilated area.

Note: This compound can damage the ozone layer if released to the atmosphere.

SAFETY GLASSES

Hexachlorobutadiene

[87–68–3]

C_4Cl_6 **UNCLASSIFIED**

HAZARDS

Liquid Can be readily absorbed through the skin. Repeated contact can cause skin and eye irritation.

Vapor Can cause kidney damage, breathing difficulty, and eye and nose irritation. May cause cancer of the kidney.

TLV/PEL 0.02 ppm

VP 0.3 mm at 25°C

AT 1130°F

Important Hazard Suspected carcinogen.

CANCER CAUSING

PRECAUTIONS

Hexachlorobutadiene is present in the residues of some plasma aluminum etchers. Etcher residues should be kept within the confines of an air-exhausted hood or used with local exhaust ventilation.

Particular care should be taken to avoid skin and eye contact. Use solvent gloves, faceshield and safety glasses, apron and armguards.

Use only within the confines of an air-exhausted hood or a closed system.

FACESHIELD AND GLASSES   SOLVENT GLOVES   APRON AND ARMGUARDS   AIR EXHAUSTED

Hexachloroethane

[67–72–1]

C_2Cl_6

UNCLASSIFIED

HAZARDS

Solid Repeated contact can cause skin and eye irritation.

Vapor Can cause eye irritation and tearing. May cause weakness, dizziness, and kidney damage. May cause cancer.

TLV/PEL	1 ppm	**FP**	385°F
IDLH	300 ppm	**AT**	743°F
VP	0.5 mm		

Important Hazard Suspected carcinogen.

CANCER CAUSING

PRECAUTIONS

Hexachloroethane is present in the residues of some plasma aluminum etchers. Etcher residues should be kept within the confines of an air-exhausted hood or used with local exhaust ventilation.

Use solvent gloves, faceshield and safety glasses, apron and armguards. Use only within the confines of an air-exhausted hood or closed system.

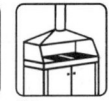

FACESHIELD AND GLASSES   SOLVENT GLOVES   APRON AND ARMGUARDS   AIR EXHAUSTED

Hexamethyldisilazane

[999–97–3]

$(CH_3)_3SiNHSi(CH_3)_3$
HMDS

FLAMMABLE

HAZARDS

Liquid Can cause burns to skin and eyes.

Vapor Ammonia-like odor detectable at low concentrations. Can be irritating to eyes, nose, throat, and lungs. Flammable.

VP 23 mm

FP 48°F

ER 0.8 – 16.3%

Important Hazard Reacts vigorously with water, alcohol, and mineral acids to give off ammonia.

FLAMMABLE SPECIAL HAZARD

(Manufacturer)

PRECAUTIONS

Avoid skin and eye contact. Use solvent gloves, faceshield and safety glasses, apron and armguards. Use only within the confines of an air-exhausted hood.
Keep away from ignition sources such as flame, spark, and heat.
Keep away from water, alcohol, and mineral acids.

FACESHIELD AND GLASSES SOLVENT GLOVES APRON AND ARMGUARDS AIR EXHAUSTED NO IGNITION SOURCE WATER REACTIVE

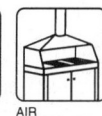

n-Hexane

[110–54–3]

$CH_3(CH_2)_4CH_3$

FLAMMABLE

HAZARDS

Liquid Can cause skin and eye irritation.

Vapor Odor detectable at 64 to 244 ppm. Prolonged exposure to 200 ppm may cause damage to the nervous system. Nausea, headache, and eye and throat irritation may occur from 10-minute exposures to 1500 ppm. 5000 ppm for 10 minutes can cause dizziness and drowsiness.

TLV/PEL	50 ppm	**FP**	−9.4°F
STEL	1000 ppm	**AT**	437°F
IDLH	1100 ppm	**ER**	1.1 – 7.5%
VP	124 mm		

Important Hazard Vapors are extremely flammable.

FLAMMABLE   SPECIAL HAZARD

PRECAUTIONS

Avoid skin and eye contact. Use solvent gloves, faceshield and safety glasses, apron and armguards. Use only within the confines of an air-exhausted hood.

Keep away from ignition sources such as flame, spark, and heat.

FACESHIELD AND GLASSES   SOLVENT GLOVES   APRON AND ARMGUARDS   AIR EXHAUSTED   NO IGNITION SOURCE

243

Hexylene Glycol

[107–41–5]

$(CH_3)_2COHCH_2CHOHCH_3$
Methyl Pentanediol

COMBUSTIBLE

HAZARDS

Liquid Can cause skin and eye irritation.

Vapor Mild sweet odor detectable at 50 ppm. This chemical is not likely to form significant vapor concentrations at room temperature (66 ppm is the maximum concentration at room temperature in a confined space). Exposure at 50 ppm can cause slight eye irritation. At 100 ppm a distinct odor is present, slight nose irritation can occur, and there may be slight discomfort during breathing. Combustible.

TLV/PEL	25 ppm C	**AT**	797°F
VP	0.05 mm	**ER**	1.3 – 7.5%
FP	200°F		

COMBUSTIBLE

PRECAUTIONS

Avoid skin and eye contact. Use solvent gloves, faceshield and safety glasses, apron and armguards. If the solution is heated, use within the confines of an air-exhausted hood.
Keep away from open flame and high temperatures.

FACESHIELD AND GLASSES	SOLVENT GLOVES	APRON AND ARMGUARDS	AIR EXHAUSTED	NO IGNITION SOURCE

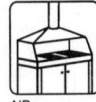

Hydrazine

H_2NNH_2

[302–01–2]

FLAMMABLE/CORROSIVE/POISON

HAZARDS

Liquid Can be readily absorbed through the skin. Can cause burns to skin and eyes. Repeated contact can cause irritation and skin rash.

Vapor Ammonia-like odor detectable at 3 to 4 ppm. Can be irritating to skin, eyes, and lungs. Can cause lung, liver, and kidney damage and convulsions. Can damage blood cells. Flammable.

TLV/PEL 0.01 ppm	**VP** 10 mm	**AT** 74 – 518°F
IDLH 50 ppm	**FP** 100°F	**ER** 2.9 – 100%

Important Hazard Suspected carcinogen. Powerful reducing agent; liquid and vapors can be explosive. May be toxic at concentrations below the odor threshold.

FLAMMABLE   CORROSIVE   POISON   CANCER CAUSING   SPECIAL HAZARD

PRECAUTIONS

Particular care should be taken to avoid skin and eye contact. Use solvent gloves, faceshield and safety glasses, apron and armguards. Use only within the confines of an air-exhausted hood.

Keep away from ignition sources such as flame, spark, or heat. Keep away from all oxidizers (including rust).

FACESHIELD AND GLASSES   SOLVENT GLOVES   APRON AND ARMGUARDS   AIR EXHAUSTED   NO IGNITION SOURCE

Hydrazine Sulfate

[10034–93–2]

$(NH_2)_2 \cdot H_2SO_4$

CORROSIVE/POISON

HAZARDS

Dust/Mist Repeated contact can cause irritation and skin rash. May cause liver damage.

Important Hazard Suspected carcinogen.

CORROSIVE   CANCER CAUSING

(Manufacturer)

PRECAUTIONS

Avoid skin and eye contact. Use solvent gloves, faceshield and safety glasses, apron and armguards. Use only within the confines of an air-exhausted hood.

FACESHIELD AND GLASSES   ACID GLOVES   APRON AND ARMGUARDS   AIR EXHAUSTED

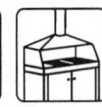

Hydrogen

[1333–74–0]

H_2

FLAMMABLE

HAZARDS

Gas Odorless. Acts as a simple asphyxiant by displacing oxygen in air. Skin contact with liquid hydrogen or cold gas vapor can cause burns and freeze tissues. Flammable.

AT 752°F

ER 4.0 – 75%

Important Hazard Can be explosive.

| FLAMMABLE | SPECIAL HAZARD | | (Liquefied Gas) 3/4/0 | (Compressed Gas) 0/4/0 |

PRECAUTIONS

Use heavy leather, cotton, or cryogenic gloves when handling liquid. Use only in a closed system or an air-exhausted hood. For some operations, use only with a combustible gas monitor present and operating.
Keep away from all ignition sources such as flame, spark, and heat.

SAFETY GLASSES   AIR EXHAUSTED   NO IGNITION SOURCE

247

Hydrogen Bromide [10035–10–6]

HBr
Hydrobromic Acid

CORROSIVE (ACID)

HAZARDS

Gas Colorless with an irritating, sharp odor detectable at 2 ppm. Inhalation of 3 to 6 ppm can cause nose and throat irritation. High concentrations can cause chemical burns to skin, eyes, and lungs. The maximum concentration that can be tolerated for 60 minutes is in the range of 50 to 100 ppm and the maximum tolerated for several hours is 10 to 50 ppm. Concentrations of 1300 to 2000 ppm can cause death on brief exposures (up to a few minutes).

TLV/PEL 3 ppm C

IDLH 30 ppm

CORROSIVE   GAS HAZARD

(Manufacturer)

PRECAUTIONS

Avoid skin and eye contact. Use only within the confines of an air-exhausted hood or closed system.

SAFETY GLASSES   AIR EXHAUSTED

Hydrogen Chloride [7647–01–0]

HCl
Hydrochloric Acid

CORROSIVE (ACID)

HAZARDS

Liquid Can cause burns to skin and eyes. Repeated or prolonged exposure to dilute solutions can cause skin rash.

Gas/Vapor Sharp odor detectable at 1 to 5 ppm. Exposure to gas can cause immediate eye irritation, coughing, burning of the throat, and a choking sensation. High concentrations can cause chemical burns to skin, eyes, and lungs. 35 ppm causes irritation of the throat. The maximum concentration that can be tolerated for 60 minutes is in the range of 50 to 100 ppm, and the maximum tolerated for several hours is 10 to 50 ppm. Concentrations of 1300 to 2000 ppm can cause death on brief exposures (up to a few minutes).

TLV/PEL 5 ppm C
IDLH 50 ppm

CORROSIVE   GAS HAZARD

PRECAUTIONS

Avoid skin and eye contact. Use acid gloves, faceshield and safety glasses, apron and armguards. Use only within the confines of an air-exhausted hood.

FACESHIELD AND GLASSES   ACID GLOVES   APRON AND ARMGUARDS   AIR EXHAUSTED

Hydrogen Cyanide [74–90–8]

HCN
Hydrocyanic Acid

**POISON/FLAMMABLE/
CORROSIVE (ACID)**

HAZARDS

Gas Bitter almond odor detectable at 0.2 to 5 ppm (40–60% of the population is not able to smell it). Gas can be absorbed through the skin in significant quantities. Short exposures below 35 ppm can cause throat irritation. Exposures above 50 ppm can cause immediate coughing, burning of the throat, and a choking sensation. Weakness, headache, confusion, nausea, and vomiting can also occur. Exposures above 90 ppm can cause death. Flammable.

TLV/PEL	10 ppm C	**IDLH**	50 ppm	**AT**	1000°F
STEL	4.7 ppm	**FP**	0°F	**ER**	5.6–40%

Important Hazard Highly poisonous. Easily absorbed through the skin. Reacts violently with acetaldehyde.

POISON — FLAMMABLE — CORROSIVE — GAS HAZARD — SPECIAL HAZARD

NFPA: 4 / 4 / 2

PRECAUTIONS

Hydrogen cyanide is formed when cyanide salts are mixed with acids. Keep cyanide salts and acids separated. If they do accidentally mix, leave the area immediately and call the emergency number. For some operations, use cyanide salts only with a hydrogen cyanide gas monitor present and operating.

Use acid gloves and safety glasses.

Cyanide salts must be used in a closed system or within an air-exhausted hood in the event hydrogen cyanide is formed.

Keep away from ignition sources such as flame, spark, and heat.

ACID GLOVES — SAFETY GLASSES — AIR EXHAUSTED — NO IGNITION SOURCE

Hydrogen Fluoride

[7664–39–3]

HF
Hydrofluoric Acid

CORROSIVE (ACID)

HAZARDS

Liquid Can cause severe burns to skin and eyes. Must not be stored in glass or metal containers.

Gas/Vapor Sharp, penetrating odor detectable at 0.04 to 0.13 ppm. 2.6 to 4.8 ppm for long periods of time may cause slight irritation of the nose, eyes, and skin. Higher concentrations can cause chemical burns to skin, eyes, and lungs. Concentrations of 50 to 250 ppm are dangerous even for brief exposures.

TLV/PEL 3 ppm C   **IDLH** 30 ppm

Important Hazard Skin and eye contact may not be immediately evident; pain may not start for up to 24 hours.

CORROSIVE	GAS HAZARD	SPECIAL HAZARD

0 / 4 / 1

PRECAUTIONS

Particular care should be taken to avoid skin and eye contact. Use acid gloves, faceshield and safety glasses, apron and armguards. Use only within the confines of an air-exhausted hood or closed system.

Use polyethylene containers.

Handle very carefully. Flush areas suspected of contact with copious amounts of water. Report all suspected exposures to the medical department after flushing the affected area.

FACESHIELD AND GLASSES	ACID GLOVES	APRON AND ARMGUARDS	AIR EXHAUSTED

Hydrogen Peroxide

[7722–84–1]

H_2O_2
Hydrogen Dioxide

OXIDIZER

HAZARDS

Liquid Can cause irritation and chemical burns to skin and eyes. Slowly decomposes in storage.

Vapor Can be irritating to skin, eyes, and lungs. High concentrations can cause severe irritation of the nose and throat and fluid in the lungs. Can cause bleaching of the hair.

TLV/PEL 1 ppm

IDLH 75 ppm

VP 5 mm at 30°C

Important Hazard Strong oxidizer.

OXIDIZER

(> 60%)

PRECAUTIONS

Avoid skin and eye contact. Use acid gloves, faceshield and safety glasses, apron and armguards. Use only within the confines of an air-exhausted hood. Store in bottles with vented caps. Keep covered.

When mixing hydrogen peroxide with sulfuric acid, add the hydrogen peroxide in small amounts to the sulfuric acid while stirring to minimize the amount of heat generated.

FACESHIELD AND GLASSES   ACID GLOVES   APRON AND ARMGUARDS   AIR EXHAUSTED

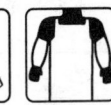

Hydrogen Selenide

[7783-07-5]

H$_2$Se

POISON/FLAMMABLE

HAZARDS

Gas Disagreeable odor detectable at 0.3 ppm. Less than 0.2 ppm can cause nausea, vomiting, metallic taste, bad breath, weakness, and fatigue. Exposure to 1.5 ppm can cause severe eye and nose irritation and fluid in the lungs (onset may be delayed for several hours). Flammable gas.

TLV/PEL 0.05 ppm

IDLH 1 ppm

Important Hazard Hydrogen selenide may be toxic at concentrations below its odor threshold.

| POISON | FLAMMABLE | GAS HAZARD | SPECIAL HAZARD |

PRECAUTIONS

Use only within a closed, well-exhausted system. Use only with a toxic gas monitor present and operating.

Keep away from ignition sources such as flame, spark, and heat.

| SAFETY GLASSES | AIR EXHAUSTED | NO IGNITION SOURCE |

253

Hydrogen Sulfide [11144–15–3]

H_2S **POISON/FLAMMABLE**

HAZARDS

Gas Rotten egg odor detectable at 0.004 to 0.13 ppm. Exposure to 50 ppm for a long period of time can cause irritation of the eyes, nose, and throat. Prolonged exposure may cause runny nose, cough, hoarseness, shortness of breath, and fluid in the lungs. 200 ppm and above can cause severe irritation, headache, nausea, vomiting, and dizziness. At 500 ppm, excitement, headache, dizziness, staggering, unconsciousness, and stopping of breathing occur in 5 minutes to 1 hour. 1000 ppm can cause death immediately. Flammable.

TLV/PEL	10 ppm	**AT**	500°F
STEL	15 ppm	**ER**	4 – 46%
IDLH	100 ppm		

Important Hazard The odor is offensive, but it is unreliable as a warning signal because the sense of smell is immediately lost at concentrations of greater than 200 ppm.

POISON   FLAMMABLE   GAS HAZARD   SPECIAL HAZARD

4 / 4 / 0

PRECAUTIONS

Use in a closed system or within an air-exhausted hood.
Keep away from ignition sources such as flame, spark, and heat.

SAFETY GLASSES   AIR EXHAUSTED   NO IGNITION SOURCE

Hydroquinone

[123–31–9]

$C_6H_4(OH)_2$
Benzohydroquinone

UNCLASSIFIED

HAZARDS

Liquid Can cause irritation of skin and eyes.

Mist/Vapor Can cause eye damage. High concentrations can cause headache, dizziness, and nausea. This chemical is not likely to form significant vapor concentrations at room temperature (0.024 ppm is the maximum concentration at 25°C in a confined space).

TLV/PEL	2 mg/m^3	**FP**	329°F
IDLH	50 mg/m^3	**AT**	930°F
VP	0.000018 mm @ 25°C		

Important Hazard Repeated skin contact may cause an allergic skin sensitization in some persons that results in severe swelling and skin reddening.

SPECIAL HAZARD

PRECAUTIONS

Avoid skin and eye contact. Use solvent gloves, faceshield and safety glasses, apron and armguards.
Use in an air-exhausted hood.

FACESHIELD AND GLASSES   SOLVENT GLOVES   APRON AND ARMGUARDS   AIR EXHAUSTED

Indium

[7440–74–6]

In **UNCLASSIFIED**

HAZARDS

Dust May cause skin and eye irritation. May deposit in the lungs and cause lung damage.

TLV/PEL 0.1 mg/m^3

(Manufacturer)

PRECAUTIONS

Avoid skin and eye contact. Use solvent or acid gloves and safety glasses. Use in an air-exhausted hood.

SAFETY GLASSES   ACID GLOVES   AIR EXHAUSTED

Indium Phosphide

[22398–80–7]

Indium Monophosphide
InP

UNCLASSIFIED

HAZARDS

Dust May cause skin and eye irritation. Indium (In) is unlikely to accumulate in the body following InP exposure.

TLV/PEL 0.1 mg/m^3 (as In)

Important Hazard Traces of phosphine can be generated when cutting or grinding solid InP with an aqueous media.

SPECIAL HAZARD

PRECAUTIONS

Avoid skin and eye contact. Use solvent or acid gloves and safety glasses. Use in an air-exhausted hood.

SAFETY GLASSES   ACID GLOVES   AIR EXHAUSTED

Inert Gases

Argon
Helium
Krypton
Nitrogen
Xenon

NONFLAMMABLE GASES

HAZARDS

Liquid Any liquefied gas freezes skin on contact.

Gas Can displace oxygen in air causing headache and dizziness at extremely high concentrations (acts as a simple asphyxiant).

3 0 0
(as Liquefied Nitrogen)

PRECAUTIONS

Use heavy leather or cotton gloves when handling liquefied gases.
Use in a well-ventilated area.

SAFETY GLASSES

Iodine

[7553–56–2]

I_2

UNCLASSIFIED

HAZARDS

Solid/Liquid Can cause severe skin and eye irritation. It is not easily removed from the skin.

Vapor Distinctive odor detectable at low concentrations. Vapor is a severe irritant (more irritating than the vapors of bromine or chlorine). Work is difficult at 0.15 to 0.20 ppm. 1 ppm can be irritating to nose and throat. 1.63 ppm can cause eye irritation. Higher concentrations can cause brown stains of the eyes and loss of outer tissue layer. May cause fluid to form in the lungs.

TLV/PEL 0.1 ppm C

IDLH 2 ppm

VP 0.3 mm

Important Hazard Can cause an allergic skin rash in certain sensitive individuals.

SPECIAL HAZARD | OXIDIZER | (Manufacturer)

PRECAUTIONS

Avoid skin and eye contact. Use solvent or acid gloves, faceshield and safety glasses, apron and armguards. Use only within the confines of an air-exhausted hood.

FACESHIELD AND GLASSES | SOLVENT GLOVES | APRON AND ARMGUARDS | AIR EXHAUSTED

Iron Oxide

[1309–37–1]

Fe_2O_3
Ferric Oxide

UNCLASSIFIED

HAZARDS

Dust May cause skin and eye irritation. May deposit in the lungs but is not considered to cause lung damage.

TLV/PEL 5 mg/m^3

IDLH 2500 mg/m^3 (as iron)

PRECAUTIONS

Avoid skin and eye contact. Use solvent or acid gloves and safety glasses.

SAFETY GLASSES   ACID GLOVES

Isobutylene

[115–11–7]

CH$_2$C(CH$_3$)CH$_3$
Isobutene
2-Methylpropene

FLAMMABLE

HAZARDS

Gas Unpleasant odor detectable at 0.6 to 1.3 ppm. Acts as a simple asphyxiant by displacing oxygen in air. May cause headache and dizziness at high concentrations. Flammable.

AT 869°F

ER 1.8 – 9.6%

FLAMMABLE

PRECAUTIONS

Use in a well-ventilated area.
Keep away from ignition sources such as flame, spark, and heat.

SAFETY GLASSES

NO IGNITION SOURCE

2-Isopropoxyethanol [109–59–1]

(CH$_3$)$_2$CHOCH$_2$CH$_2$OH
Ethylene Glycol Monopropyl Ether

FLAMMABLE

HAZARDS

Liquid Can be readily absorbed through the skin. Repeated contact can cause skin and eye irritation.

Vapor Can be irritating to skin, eyes, and lungs. May cause lung and blood damage. High concentrations can cause headache, dizziness, drowsiness, and difficulty in breathing. Flammable.

TLV	25 ppm
VP	2.6 mm
FP	92°F

FLAMMABLE

1-3-0

PRECAUTIONS

Particular care should be taken to avoid skin and eye contact. Use solvent gloves, faceshield and safety glasses, apron and armguards. Use only within the confines of an air-exhausted hood.

Keep away from ignition sources such as flame, spark, and heat.

FACESHIELD AND GLASSES | SOLVENT GLOVES | APRON AND ARMGUARDS | AIR EXHAUSTED | NO IGNITION SOURCE

Isopropyl Alcohol

[67–63–0]

C_3H_8O
Isopropanol
IPA

FLAMMABLE

HAZARDS

Liquid Repeated contact can cause skin and eye irritation.

Vapor Fragrant odor detectable at 3 to 28 ppm. Mild irritation of eyes, nose, and throat occurs at 400 ppm. Inhalation of higher concentrations can cause flushing, headache, dizziness, mental depression, nausea, vomiting, narcosis, anesthesia, and coma. Flammable.

TLV/PEL	400 ppm	**VP**	33 mm
STEL	500 ppm	**FP**	53°F
IDLH	2000 ppm	**AT**	750°F
ER	2 – 12%		

FLAMMABLE

3/1/0

PRECAUTIONS

Avoid skin and eye contact. Use solvent gloves, faceshield and safety glasses, apron and armguards. Use only within the confines of an air-exhausted hood.

Keep away from ignition sources such as flame, spark, and heat.

FACESHIELD AND GLASSES | SOLVENT GLOVES | APRON AND ARMGUARDS | AIR EXHAUSTED | NO IGNITION SOURCE

Kerosene

[8008–20–6]

Coal Oil
Kerosine

COMBUSTIBLE

HAZARDS

Liquid Repeated contact can cause skin irritation. Can cause eye irritation.

Vapor Mild petroleum odor detectable at 1 ppm. Inhalation of high concentrations can cause headaches and dizziness. Combustible.

TLV/PEL	14 ppm (NIOSH)		
VP	5 mm (approx.)	**AT**	410°F
FP	120°F	**ER**	0.7 – 5%

COMBUSTIBLE

PRECAUTIONS

Avoid skin and eye contact. Use solvent gloves, faceshield and safety glasses, apron and armguards. Use only within the confines of an air-exhausted hood.

Keep away from open flame and high temperatures.

FACESHIELD AND GLASSES | SOLVENT GLOVES | APRON AND ARMGUARDS | AIR EXHAUSTED | NO IGNITION SOURCE

Lead

[7439–92–1]

Pb **POISON**

HAZARDS

Dust Inhalation and ingestion of lead compounds can damage the blood cells and nervous system. Can cause headache, insomnia, dizziness, and anemia. Can cause skin and eye irritation.

TLV/PEL 0.05 mg/m^3

IDLH 100 mg/m^3

Important Hazard Cumulative poison. Can cause reproductive disorders. Suspected carcinogen. Emits toxic fumes when heated above 800°F.

| POISON | SPECIAL HAZARD | CANCER CAUSING | (Manufacturer) |

PRECAUTIONS

Avoid inhalation and eye and skin contact. Use acid or solvent gloves and safety glasses.

The brushing, sweeping, or scraping of lead must be done in an air-exhausted hood or with local exhaust.

Wash hands thoroughly after handling and before eating, drinking, or smoking.

ACID GLOVES   SAFETY GLASSES   AIR EXHAUSTED

Lead Acetate

[301–04–02]

$(CH_3COO)_2Pb \cdot 2H_2O$

POISON

HAZARDS

Dust Inhalation and ingestion of lead compounds can damage the blood cells and nervous system. Can cause headache, insomnia, dizziness and anemia. Can cause skin and eye irritation.

TLV/PEL 0.05 mg/m^3 (as lead)

Important Hazard Cumulative poison. Can cause reproductive disorders. Suspected carcinogen.

| POISON | SPECIAL HAZARD | CANCER CAUSING | (Manufacturer) 4/0/0 |

PRECAUTIONS

Avoid inhalation and eye and skin contact. Use acid or solvent gloves and safety glasses. Use in a well-ventilated area.

The brushing, sweeping, or scraping of lead acetate must be done in an air-exhausted hood or with local exhaust.

Wash hands thoroughly after handling and before eating, drinking, or smoking.

ACID GLOVES | SAFETY GLASSES | AIR EXHAUSTED

Lead Nitrate

[10099–74–8]

$Pb(NO_3)_2$

POISON/OXIDIZER

HAZARDS

Dust Inhalation and ingestion of lead compounds can damage the blood cells and nervous system. Can cause headache, insomnia, dizziness, and anemia. Can cause skin and eye irritation.

TLV/PEL 0.05 mg/m^3 (as lead)

Important Hazard Cumulative poison. Can cause reproductive disorders. Suspected carcinogen.

POISON OXIDIZER SPECIAL HAZARD CANCER CAUSING

(Manufacturer)

PRECAUTIONS

Avoid inhalation and eye and skin contact. Use acid or solvent gloves and safety glasses. Use in a well-ventilated area.

The brushing, sweeping, or scraping of lead nitrate must be done in an air-exhausted hood or with local exhaust.

Wash hands thoroughly after handling and before eating, drinking, or smoking.

ACID GLOVES   SAFETY GLASSES   AIR EXHAUSTED

Limonene

[0138–86–3]

COMBUSTIBLE

$C_{10}H_{16}$
d-Limonene and 1-Limonene
mixture

HAZARDS

Liquid Can cause skin and eye irritation. Repeated contact can cause an allergic skin rash. May be readily absorbed through the skin.

Vapor Pleasant, lemon-like odor detectable at as low as 0.001 ppm. High concentrations may cause throat and lung irritation. Combustible.

VP   1 mm @ 14°C

FP   113°F

ER   0.7 – 6.1%

COMBUSTIBLE

1 / 1 / 0 (Manufacturer)

PRECAUTIONS

Particular care should be taken to avoid skin and eye contact. Use solvent gloves, faceshield and safety glasses, apron and armguards. Use in an air-exhausted hood.
Keep away from open flame and high temperatures.

FACESHIELD AND GLASSES | SOLVENT GLOVES | APRON AND ARMGUARDS | AIR EXHAUSTED | NO IGNITION SOURCE

Liquefied Petroleum Gas [74–98–6]

LPG **FLAMMABLE GAS**

HAZARDS

Liquid Freezes skin on contact.

Gas Faint odor detectable at 12,000 to 20,000 ppm when pure (fuel grades have mercaptan odorant added that gives the propane mixture a skunk-like odor detectable at low concentrations). Acts as a simple asphyxiant by diluting oxygen in air. High concentrations can cause dizziness and headache.

TLV/PEL 1000 ppm

IDLH 2000 ppm

AT 842°F

ER 2.2 – 9.5%

Important Hazard Extremely flammable.

FLAMMABLE SPECIAL HAZARD

4 / 1 / 0

PRECAUTIONS

Avoid all direct contact. Leather gloves must be worn when filling tanks.
Use only within a closed system.
Keep away from strong oxidizers and ignition sources such as flame, spark, and heat.

SAFETY GLASSES NO IGNITION SOURCE AIR EXHAUSTED

Lithium Hydroxide [1310–65–2]

LiOH **CORROSIVE (BASE)**

HAZARDS

Solid/Liquid Can cause severe burns to skin and eyes.

Dust/Mist Can be highly irritating to skin, eyes, nose, and throat (similar to other strong bases such as sodium hydroxide and potassium hydroxide). Exposure to high concentrations may cause swelling of airways, which can make breathing difficult and could be fatal.

CORROSIVE

PRECAUTIONS

Avoid skin and eye contact. Use acid gloves, faceshield and safety glasses, apron and armguards. Use only within the confines of an air-exhausted hood.

FACESHIELD AND GLASSES   ACID GLOVES   APRON AND ARMGUARDS   AIR EXHAUSTED

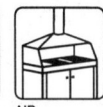

Magnesium [7439–95–4]

Mg **FLAMMABLE SOLID**

HAZARDS

Solid Repeated contact can cause skin irritation. Small particles imbedded in the skin may cause sores which may become infected.

Dust/Fume Exposure to fume can cause metal fume fever — fever, cough, nausea, chest pain, increase in white blood cells. High concentrations of dust can cause irritation of eyes and nose.

AT 883°F

Important Hazard Fire hazard in the form of dust or flakes. Magnesium fires do not flare up violently unless moisture is present. Magnesium reacts with moisture and acids to form hydrogen, which is a highly dangerous fire and explosion hazard. If fire occurs, smother with inert material or use approved Class D extinguishers; DO NOT use carbon dioxide or halogenated extinguishers; DO NOT use water or foam.

FLAMMABLE   SPECIAL HAZARD

PRECAUTIONS

Avoid skin and eye contact. Use acid or solvent gloves and safety glasses. Use in a well-ventilated area.
Keep away from moisture, including small amounts of water.
Keep dust and flakes away from ignition sources such as flame, spark, and heat.

SAFETY GLASSES   ACID GLOVES   NO IGNITION SOURCE   WATER REACTIVE

Magnesium Acetate

[142–72–3]

$C_4H_6O_4Mg$
Acetic Acid
Magnesium Salt

UNCLASSIFIED

HAZARDS

Dust May cause skin, eye, nose and throat irritation and nasal discharge and coughing.

PRECAUTIONS

Avoid skin and eye contact. Use acid or solvent gloves and safety glasses. Use in a well-ventilated area.

SAFETY GLASSES

ACID GLOVES

Magnesium Chloride

[7786-30-3]

MgCl$_2$

UNCLASSIFIED

HAZARDS

Dust May cause skin and eye irritation. High concentrations of dust can cause irritation of eyes and nose.

(Manufacturer)

PRECAUTIONS

Avoid skin and eye contact. Use acid or solvent gloves and safety glasses. Use in a well-ventilated area.

SAFETY GLASSES

ACID GLOVES

Magnesium Dioxide

[1335–26–8]

MgO$_2$
Magnesium Peroxide

UNCLASSIFIED

HAZARDS

Dust May cause skin and eye irritation. High concentrations may cause nausea, weakness, chills, and loss of appetite. High concentrations of dust can cause irritation of eyes and nose.

PRECAUTIONS

Avoid skin and eye contact. Use acid or solvent gloves and safety glasses. Use in a well-ventilated area.

SAFETY GLASSES

ACID GLOVES

Magnesium Oxide

[1309–48–4]

MgO
Calcined Magnesia

UNCLASSIFIED

HAZARDS

Solid   Repeated contact can cause skin irritation.

Dust/Fume   Exposure to fume can cause metal fume fever, cough, nausea, chest pain, increase in white blood cells. High concentrations of dust can cause irritation of eyes and nose.

TLV/PEL   10 mg/m^3 (as fume)

(Manufacturer)

PRECAUTIONS

Avoid skin and eye contact. Use acid or solvent gloves and safety glasses. Use in a well-ventilated area.

SAFETY GLASSES

ACID GLOVES

Manganous Sulfate

[7785–87–7]

MnSO$_4$
Manganese Sulfate

UNCLASSIFIED

HAZARDS

Dust Can cause skin and eye irritation. May cause nausea, chills, weakness, and loss of appetite. May cause tiredness, sleepiness, and weakness in the legs.

TLV/PEL 0.2 mg/m^3 C (as manganese)

IDLH 500 mg/m^3 (as manganese)

(Manufacturer)

PRECAUTIONS

Avoid skin and eye contact. Use acid or solvent gloves and safety glasses. Use in a well-ventilated area.

SAFETY GLASSES   ACID GLOVES

Melamine

[108–78–1]

(NC)$_3$(NH$_2$)$_3$

UNCLASSIFIED

HAZARDS

Solid Repeated contact may cause skin and eye irritation.

Dust Can cause skin and eye irritation.

TLV/PEL 10 mg/m^3 (as a nuisance dust)

VP 50 mm at 315°C

Important Hazard When heated above 375°C emits highly toxic fumes of cyanides and oxides of nitrogen.

SPECIAL HAZARD

(Manufacturer)

PRECAUTIONS

Avoid skin and eye contact. Use solvent or acid gloves and safety glasses.
Avoid excess heat without local exhaust.
Use in well-ventilated area.

SAFETY GLASSES   SOLVENT GLOVES

Mercuric Acetate

[1600–27–7]

$Hg(C_2H_3O_2)_2$
Mercury Acetate

POISON

HAZARDS

Solid/Liquid Can cause skin and eye irritation. Repeated exposure can cause skin rash. Can be readily absorbed through the skin.

Dust/Mist Acute exposure can cause coughing, chest pain, headache, and breathing difficulties. Chronic exposure can damage the nervous system and kidneys.

TLV/PEL 0.01 mg/m^3 (as alkyl mercury compound)

STEL 0.03 mg/m^3 (as alkyl mercury compound)

IDLH 2 mg/m^3 (as mercury)

POISON

PRECAUTIONS

Particular care should be taken to avoid skin and eye contact. Use solvent gloves, faceshield and safety glasses, apron and armguards. Use only within the confines of an air-exhausted hood.

FACESHIELD AND GLASSES | SOLVENT GLOVES | APRON AND ARMGUARDS | AIR EXHAUSTED

Mercuric Oxide

[21908-53-2]

HgO

POISON

HAZARDS

Dust May cause severe skin and eye irritation. Can be readily absorbed through the skin. Acute exposure can cause coughing, chest pain, headache, and difficulty in breathing. Chronic exposure can cause difficulty in sleeping, tremors, weakness, loss of appetite, and kidney, lung, and gum/mouth disorders.

TLV/PEL 0.025 mg/m^3

IDLH 10 mg/m^3 (as mercury)

POISON

PRECAUTIONS

Particular care should be taken to avoid skin and eye contact. Use acid or solvent gloves and safety glasses.
Use only within the confines of an air-exhausted hood.

SAFETY GLASSES   ACID GLOVES   AIR EXHAUSTED

Mercury, Metal

[7439-97-6]

Hg
Quicksilver

POISON

HAZARDS

Liquid Can cause skin and eye irritation. Repeated exposure can cause skin rash. Can be readily absorbed through the skin. Evaporates at room temperature.

Vapor Exposure to 1.2 to 8.5 mg/m^3 can cause coughing, chest pain, headache, and difficulty in breathing. Repeated exposures can cause difficulty in sleeping, tremors, weakness, loss of appetite, and kidney, lung, and gum/mouth disorders.

TLV/PEL 0.025 mg/m^3 **VP** 0.0012 mm

IDLH 10 mg/m^3

Important Hazard Emits highly toxic fumes when heated.

POISON SPECIAL HAZARD

(Manufacturer)

PRECAUTIONS

Particular care should be taken to avoid skin and eye contact. Use acid or solvent gloves and safety glasses.

Use only within the confines of an air-exhausted hood.

Mercury from ruptured mercury arc lamps or broken thermometers should be cleaned up immediately and stored in a sealed container.

SAFETY GLASSES ACID GLOVES AIR EXHAUSTED

280

Methane

[74–82–8]

CH$_4$ **FLAMMABLE**

HAZARDS

Gas Odorless or possibly slight odor at high concentrations. Acts as a simple asphyxiant by displacing oxygen in air. Flammable.

AT 999°F

ER 5 – 15.4%

FLAMMABLE

PRECAUTIONS

Use in a well-ventilated area.
Keep away from ignition sources such as flame, spark, and heat.

SAFETY GLASSES

NO IGNITION SOURCE

2-Methoxyethanol*

[109–86–4]

CH$_3$OCH$_2$CH$_2$OH **COMBUSTIBLE**
Ethylene Glycol Monomethyl Ether

HAZARDS

Liquid Can be readily absorbed through the skin. Repeated contact can cause skin and eye irritation.

Vapor Sweet odor detectable at 0.09 to 60 ppm. Repeated exposure may cause damage to the blood cells, liver, and kidneys. High concentrations may cause headache, confusion, agitation, disorientation, weakness, nausea, breathing difficulties, and increased heart rate. Combustible.

TLV/PEL	5 ppm	**FP** 102°F	**VP** 6 mm
IDLH	200 ppm	**AT** 545°F	**ER** 2.5 – 14.8%

Important Hazard Suspected of causing reproductive disorders.

COMBUSTIBLE SPECIAL HAZARD

PRECAUTIONS

Particular care should be taken to avoid skin and eye contact. Use solvent gloves, faceshield and safety glasses, apron and armguards. Use only within the confines of an air-exhausted hood.

Keep away from open flame and high temperatures.

FACESHIELD AND GLASSES SOLVENT GLOVES APRON AND ARMGUARDS AIR EXHAUSTED NO IGNITION SOURCE

* The Semiconductor Industry Association recommends that its member companies eliminate the use of this chemical from their manufacturing operations.

2-Methoxyethyl Acetate* [110–49–6]

CH$_3$COOCH$_2$CH$_2$OCH$_3$ **COMBUSTIBLE**
Ethylene Glycol Monomethyl Ether
  Acetate

HAZARDS

Liquid Can be readily absorbed through the skin. Can cause eye irritation.

Vapor Sweet odor detectable at 0.3 to 0.6 ppm. Repeated exposure may cause damage to the blood cells and kidneys. High concentration may cause headache, confusion, agitation, disorientation, weakness, nausea, breathing difficulties, and increased heart rate. Combustible.

TLV/PEL	5 ppm	**VP**	2 mm	**AT**	740°F
IDLH	200 ppm	**FP**	111°F	**ER**	1.1 – 8.2%

Important Hazard Suspected of causing reproductive disorders.

COMBUSTIBLE SPECIAL HAZARD

PRECAUTIONS

Particular care should be taken to avoid skin and eye contact. Use solvent gloves, faceshield and safety glasses, apron and armguards. Use only within the confines of an air-exhausted hood.
Keep away from open flame and high temperatures.

FACESHIELD AND GLASSES SOLVENT GLOVES APRON AND ARMGUARDS AIR EXHAUSTED NO IGNITION SOURCE

* The Semiconductor Industry Association recommends that its member companies eliminate the use of this chemical from their manufacturing operations.

1-Methoxy-2-Propyl Acetate

[108–65–6]

$C_6H_{12}O_3$
Propylene Glycol Monomethyl
  Ether Acetate
PGMEA

COMBUSTIBLE

HAZARDS

Vapor May cause irritation of nose, throat, and lungs. High concentrations may cause headache, nausea, vomiting, diarrhea, dizziness, drowsiness, and incoordination. Combustible.

VP 3.7 mm

FP 108°F

ER 1.3 – 13.1%

COMBUSTIBLE

0 / 2 / 0

PRECAUTIONS

Avoid skin and eye contact. Use solvent gloves, faceshield and safety glasses, apron and armguards. Use in an air-exhausted hood.
Keep away from open flame and high temperatures.

| FACESHIELD AND GLASSES | SOLVENT GLOVES | APRON AND ARMGUARDS | AIR EXHAUSTED | NO IGNITION SOURCE |

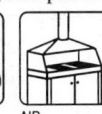

Methyl Alcohol

[67–56–1]

CH₃OH
Methanol
Wood Alcohol

FLAMMABLE

HAZARDS

Liquid Repeated contact can cause skin and eye irritation. Can be readily absorbed through the skin.

Vapor Characteristic pungent odor detectable at 4 to 6000 ppm. Exposure to 365 to 3080 ppm may cause blurred vision, headache, dizziness, and nausea. Repeated exposure to 1200 to 8300 ppm may cause damage to the nervous system and blindness. Flammable.

TLV/PEL	200 ppm	**FP**	52°F
STEL	250 ppm	**AT**	878°F
IDLH	6000 ppm	**ER**	6.7 – 36.5%
VP	97 mm		

FLAMMABLE

3 / 1 / 0

PRECAUTIONS

Particular care should be taken to avoid skin and eye contact. Use solvent gloves, faceshield and safety glasses, apron and armguards. Use only within the confines of an air-exhausted hood.

Keep away from ignition sources such as flame, spark, and heat.

FACESHIELD AND GLASSES   SOLVENT GLOVES   APRON AND ARMGUARDS   AIR EXHAUSTED   NO IGNITION SOURCE

Methyl n-Butyl Ketone [591–78–6]

$CH_3COCH_2CH_2CH_2CH_3$ **FLAMMABLE**
2-Hexanone
MBK

HAZARDS

Liquid Can be readily absorbed through the skin. Can cause eye irritation. Repeated contact may cause skin irritation.

Vapor Characteristic acetone-like odor detectable at 0.07 to 0.09 ppm. Repeated exposure above 140 ppm can cause nerve damage, weakness of the hands and legs, and tingling in the hands and feet. Exposure at 1000 ppm for several minutes can cause eye and nose irritation. Flammable.

TLV/PEL	5 ppm	**FP**	73°F
IDLH	1600 ppm	**AT**	795°F
VP	3 mm	**ER**	1.2 – 8%

FLAMMABLE

PRECAUTIONS

Particular care should be taken to avoid skin and eye contact. Use solvent gloves, faceshield and safety glasses, apron and armguards. Use only within the confines of an air-exhausted hood.
Keep away from ignition sources such as flame, spark, and heat.

FACESHIELD AND GLASSES · SOLVENT GLOVES · APRON AND ARMGUARDS · AIR EXHAUSTED · NO IGNITION SOURCE

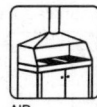

Methyl Chloride

[74–87–3]

CH₃Cl
Chloromethane

FLAMMABLE

HAZARDS

Gas Faint sweet odor detectable above 100 ppm. Repeated exposure to low concentrations can cause damage to heart and nervous system. Can cause liver, kidney, and bone marrow damage. Skin contact can produce an anesthetic effect. Exposures at 1000 ppm may cause dizziness, drowsiness, incoordination, nausea, vomiting, abdominal pains, and blurred vision. Flammable.

TLV/PEL	50 ppm	**AT**	1170°F
STEL	100 ppm	**ER**	7.6 – 17.4%
FP	– 50°F		

Important Hazard Symptoms may be delayed for several hours. Suspected carcinogen. May be toxic at concentrations below the odor threshold.

FLAMMABLE — GAS HAZARD — SPECIAL HAZARD — CANCER CAUSING

(4 / 1 / 0)

PRECAUTIONS

Avoid skin and eye contact. Use only within a closed system.
Keep away from ignition sources such as flame, spark, and heat.

SAFETY GLASSES — AIR EXHAUSTED — NO IGNITION SOURCE

Methylene Bisphenyl Isocyanate

[101–68–8]

$CH_2(C_6H_4NCO)_2$
MDI

COMBUSTIBLE

HAZARDS

Liquid Can cause skin and eye irritation.

Vapor/Mist Slightly musty odor detectable at 0.4 ppm. Can be irritating to skin, eyes, nose, throat, and lungs. Can damage nose and throat. Combustible.

TLV/PEL	0.005 ppm	**FP**	385°F
IDLH	75 mg/m^3	**AT**	464°F
VP	0.00014 mm at 25°C		

Important Hazard Low-level exposures can cause allergic sensitization of the lungs in sensitive individuals.

COMBUSTIBLE   SPECIAL HAZARD

2 / 1 / 0 (Manufacturer)

PRECAUTIONS

Avoid skin and eye contact. Use solvent gloves, faceshield and safety glasses, apron and armguards. Use in a well-ventilated area.
Keep away from open flame and high temperatures.

FACESHIELD AND GLASSES   SOLVENT GLOVES   APRON AND ARMGUARDS   NO IGNITION SOURCE

Methylene Chloride

[75-09-2]

UNCLASSIFIED

CH_2Cl_2
Dichloromethane

HAZARDS

Liquid Repeated contact can cause skin and eye irritation.

Vapor Chloroform-like odor detectable at 1 to 440 ppm. Headache and dizziness can occur from exposure at 200 ppm for 2 to 3 hours. Higher concentrations can cause slight eye, nose, and throat irritation. May cause liver and kidney damage. May cause liver cancer. It is not flammable at ordinary temperatures, but flammable vapor-air mixtures can form at about 212°F.

TLV/PEL 50 ppm   **AT** 1033°F   **VP** 350 mm

IDLH 2300 ppm   **ER** 12 – 19%

Important Hazard Emits highly toxic phosgene fumes when heated to decomposition. Suspected human carcinogen.

CANCER CAUSING   SPECIAL HAZARD

PRECAUTIONS

Avoid skin and eye contact. Use solvent gloves, faceshield and safety glasses, apron and armguards. Use only within the confines of an air-exhausted hood. Keep away from high temperatures.

Note: Particular care should be taken in the selection of solvent gloves for use with methylene chloride to ensure the gloves will not dissolve on contact with the chemical.

FACESHIELD AND GLASSES   SOLVENT GLOVES   APRON AND ARMGUARDS   AIR EXHAUSTED   NO IGNITION SOURCE

Methyl Ethyl Ketone

[78–93–3]

C_4H_8O
2-Butanone
MEK

FLAMMABLE

HAZARDS

Liquid Repeated contact can cause skin and eye irritation.

Vapor Fragrant mint-like odor detectable at 2 to 5 ppm. Exposure at 100 to 200 ppm may cause mild irritation of eyes, nose, and throat. Higher concentrations may cause headache, numbness in arms and legs, dizziness, incoordination, and, eventually, unconsciousness. Flammable.

TLV/PEL	200 ppm	**FP**	21°F
STEL	300 ppm	**AT**	960°F
IDLH	3000 ppm	**ER**	2 – 12%
VP	77.5 mm		

FLAMMABLE

PRECAUTIONS

Avoid skin and eye contact. Use solvent gloves, faceshield and safety glasses, apron and armguards. Use only within the confines of an air-exhausted hood.

Keep away from ignition sources such as flame, spark, and heat.

FACESHIELD AND GLASSES   SOLVENT GLOVES   APRON AND ARMGUARDS   AIR EXHAUSTED   NO IGNITION SOURCE

Methyl Isoamyl Ketone

[110–12–3]

FLAMMABLE

$C_7H_{14}O$
5-Methyl-2-Hexanone
MIAK

HAZARDS

Liquid Repeated contact can cause skin and eye irritation.

Vapor Pleasant odor detectable at 0.01 to 0.05 ppm. Can be irritating to eyes and lungs. Inhalation of high concentrations can cause headaches and dizziness. Combustible.

TLV/PEL	50 ppm
VP	4.5 mm
FP	96°F
AT	375°F
ER	1.0 – 8.2%

FLAMMABLE

PRECAUTIONS

Avoid skin and eye contact. Use solvent gloves, faceshield and safety glasses, apron and armguards. Use only within the confines of an air-exhausted hood.

Keep away from open flame and high temperatures.

FACESHIELD AND GLASSES   SOLVENT GLOVES   APRON AND ARMGUARDS   AIR EXHAUSTED   NO IGNITION SOURCE

Methyl Isobutyl Ketone

[108–10–1]

$CH_3COCH_2CH(CH_3)_2$
Hexone
MIBK

FLAMMABLE

HAZARDS

Liquid Can cause eye irritation. Repeated contact can cause skin irritation.

Vapor Characteristic sweet pungent odor detectable at 0.1 to 8 ppm. Eye, nose, and throat irritation can occur at 200 to 400 ppm. High concentrations can cause headache, weakness, loss of appetite, upset stomach, nausea, and vomiting. Repeated exposures at high concentrations may cause liver and kidney damage. Flammable.

TLV/PEL	50 ppm	**VP**	15 mm	**ER**	1.2 – 8.0%
STEL	75 ppm	**FP**	64°F		
IDLH	500 ppm	**AT**	800°F		

FLAMMABLE

PRECAUTIONS

Avoid skin and eye contact. Use solvent gloves, faceshield and safety glasses, apron and armguards. Use only within the confines of an air-exhausted hood.

Keep away from ignition sources such as flame, spark, and heat.

FACESHIELD AND GLASSES SOLVENT GLOVES APRON AND ARMGUARDS AIR EXHAUSTED NO IGNITION SOURCE

Methyl Isopropyl Ketone

[563–80–4]

CH$_3$COCH(CH$_3$)$_2$
3-Methyl-2-Butanone
MIPK

FLAMMABLE

HAZARDS

Liquid May cause eye irritation. Repeated contact may cause skin irritation.

Vapor Odor detectable at 4 to 5 ppm. May cause irritation of the eyes and nose. High concentrations may cause dizziness. Flammable.

TLV/PEL 200 ppm

FP 91°F

VP 10 mm at 8.3°C

ER 1.3 – 8.2%

FLAMMABLE

3 / 2 / 0 (Manufacturer)

PRECAUTIONS

Avoid skin and eye contact. Use solvent gloves, faceshield and safety glasses, apron and armguards. Use only within the confines of an air-exhausted hood.
Keep away from ignition sources such as flame, spark, and heat.

FACESHIELD AND GLASSES | SOLVENT GLOVES | APRON AND ARMGUARDS | AIR EXHAUSTED | NO IGNITION SOURCE

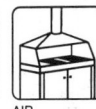

Methyl Phenyl Ether

[100–66–3]

$CH_3OC_6H_5$
Anisole
Methoxybenzene

COMBUSTIBLE

HAZARDS

Liquid Can cause skin and eye irritation.

Vapor Can cause irritation to eyes, nose, and throat. Combustible.

VP 3.1 mm at 25°C

FP 125°F

AT 887°F

COMBUSTIBLE

PRECAUTIONS

Avoid skin and eye contact. Use solvent gloves, faceshield and safety glasses, apron and armguards. Use only within the confines of an air-exhausted hood.

Keep away from open flame and high temperatures.

| FACESHIELD AND GLASSES | SOLVENT GLOVES | APRON AND ARMGUARDS | AIR EXHAUSTED | NO IGNITION SOURCE |

Methyl Propyl Ketone

[107–87–9]

$CH_3COC_3H_7$
2-Pentanone
MPK

FLAMMABLE

HAZARDS

Liquid Can cause eye irritation. Repeated contact can cause skin irritation.

Vapor Odor detectable at 3 to 14 ppm. 1500 ppm can cause irritation of the eyes and nose. High concentrations may cause dizziness. Flammable.

TLV/PEL	200 ppm	**FP**	45°F
STEL	250 ppm	**AT**	846°F
IDLH	5000 ppm	**ER**	1.5 – 8.2%
VP	27 mm		

FLAMMABLE

PRECAUTIONS

Avoid skin and eye contact. Use solvent gloves, faceshield and safety glasses, apron and armguards. Use only within the confines of an air-exhausted hood.

Keep away from ignition sources such as flame, spark, and heat.

FACESHIELD AND GLASSES    SOLVENT GLOVES    APRON AND ARMGUARDS    AIR EXHAUSTED    NO IGNITION SOURCE

Mineral Oil

[8012–95–1]

C_{10} to C_{14} Hydrocarbons
Paraffin Oil
Oil Mist, Mineral

UNCLASSIFIED

HAZARDS

Liquid Repeated contact may cause skin irritation. May cause eye irritation.

Mist May cause minor irritation of the lungs. Combustible.

TLV/PEL 5 mg/m^3 (severely refined)
0.2 mg/m^3 (mildly refined)

IDLH 2500 mg/m^3

FP 275 – 444°F

Important Hazard Mildly refined mineral oil is a confirmed human carcinogen.

COMBUSTIBLE CANCER CAUSING

PRECAUTIONS

Avoid skin and eye contact. Use solvent gloves and safety glasses.
Use in a well-ventilated area. Keep away from open flame and high temperatures.

SAFETY GLASSES SOLVENT GLOVES NO IGNITION SOURCE

Morpholine

[110–91–8]

C_4H_9NO
Diethylene Oximide

FLAMMABLE

HAZARDS

Liquid Can cause severe skin and eye irritation. Can be readily absorbed through the skin.

Vapor Unpleasant, fishy odor detectable at 0.01 to 0.07 ppm. Can cause skin, eye, nose, and lung irritation. May cause nausea, vomiting, headache, and vision problems. Repeated contact at high concentrations may cause eye, lung, liver, and kidney damage. Flammable.

TLV/PEL	20 ppm	**FP**	95°F
IDLH	1400 ppm	**AT**	555°F
VP	6.6 mm	**ER**	1.4 – 11.2%

FLAMMABLE

3/3/0

PRECAUTIONS

Particular care should be taken to avoid skin and eye contact. Use solvent gloves, faceshield and safety glasses, apron and armguards. Use only within the confines of an air-exhausted hood.

Keep away from ignition sources such as flame, spark, and heat.

FACESHIELD AND GLASSES • SOLVENT GLOVES • APRON AND ARMGUARDS • AIR EXHAUSTED • NO IGNITION SOURCE

M-Pyrol

[872–50–4]

C_5H_9NO
N-Methyl-2-Pyrrolidone
NMP

COMBUSTIBLE

HAZARDS

Liquid   Can be readily absorbed through the skin. Can cause severe eye irritation. Repeated or prolonged contact can cause skin irritation.

Vapor   Strong solvent odor. 1 to 2 ppm uncomfortable after about 30 minutes. High concentrations may cause headache and irritation of eyes, nose, and throat. Combustible.

TLV/PEL   100 ppm (Manufacturer)

VP	0.29 mm	**AT**	518°F
FP	128°F	**ER**	1.3 – 12%

COMBUSTIBLE

PRECAUTIONS

Avoid skin and eye contact. Use solvent gloves, faceshield and safety glasses, apron and armguards. Use only within the confines of an air-exhausted hood.

Note: Particular care should be taken in the selection of solvent gloves for use with M-pyrol to ensure the gloves will not dissolve on contact with the chemical.

Keep away from open flame and high temperatures.

FACESHIELD AND GLASSES   SOLVENT GLOVES   APRON AND ARMGUARDS   AIR EXHAUSTED   NO IGNITION SOURCE

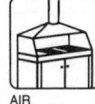

Naphtha

[8030–30–6]

Rubber Solvent
Petroleum Distillates
Petroleum Ether

FLAMMABLE

HAZARDS

Liquid Can cause skin and eye irritation.

Vapor Exposure to 430 ppm can cause eye and throat irritation. High concentrations can cause dizziness.

TLV/PEL 400 ppm

IDLH 1000 ppm

VP 40 mm (approx.)

FP –40 to –86°F

ER 1 – 6%

Important Hazard Vapors are extremely flammable.

FLAMMABLE   SPECIAL HAZARD

(Manufacturer) 2/4/0

PRECAUTIONS

Avoid skin and eye contact. Use solvent gloves, faceshield and safety glasses, apron and armguards. Use only within the confines of an air-exhausted hood.

Keep away from ignition sources such as flame, spark, and heat.

FACESHIELD AND GLASSES   SOLVENT GLOVES   APRON AND ARMGUARDS   AIR EXHAUSTED   NO IGNITION SOURCE

299

Naphthalene

[91–20–3]

$C_{10}H_8$

COMBUSTIBLE

HAZARDS

Solid/Liquid Repeated contact can cause skin and eye irritation.

Vapor Mothball odor detectable at 0.03 to 0.8 ppm. Can be irritating to eyes at 15 ppm. Can cause headache, nausea, and loss of appetite. Can cause eye, blood, liver, and kidney damage. 71 ppm is the maximum vapor concentration at room temperature in a confined space. Combustible.

TLV/PEL	10 ppm
STEL	15 ppm
IDLH	250 ppm
VP	0.054 mm
FP	174°F
AT	979°F
ER	0.9 – 5.9%

COMBUSTIBLE

PRECAUTIONS

Avoid skin and eye contact. Use solvent gloves, faceshield and safety glasses, apron and armguards. Use only within the confines of an air-exhausted hood.

Keep away from open flame and high temperatures.

FACESHIELD AND GLASSES | SOLVENT GLOVES | APRON AND ARMGUARDS | AIR EXHAUSTED | NO IGNITION SOURCE

Nickel

(elemental, insoluble, and soluble compounds)

Ni, NiBr$_2$, NiCl$_2$,
Nickel Acetate,
Nickel Sulfamate, etc.

UNCLASSIFIED

HAZARDS

Dust/Fume Can cause skin and eye irritation. Repeated skin contact can cause allergic skin reactions. Inhalation can cause metallic taste, nausea, tightness of the chest, and fever. Can also cause irritation of the eyes, nose, throat, and skin. Chronic exposure may cause nerve and lung damage, and may aggravate a preexisting lung condition or allergy in some workers.

TLV/PEL 0.05 mg/m^3

IDLH 10 mg/m^3 (as nickel)

Important Hazard Confirmed human carcinogen.

CANCER CAUSING

(Manufacturer)

PRECAUTIONS

Avoid skin and eye contact. Use acid or solvent gloves and safety glasses. Use only within the confines of an air-exhausted hood.

ACID GLOVES   SAFETY GLASSES   AIR EXHAUSTED

Nitric Acid

HNO_3

[7697–37–2]

**OXIDIZER/
CORROSIVE (ACID)**

HAZARDS

Liquid Can cause severe burns to skin and eyes. Can stain the skin bright yellow. Solutions greater than 40% are oxidizers.

Vapor Sharp odor detectable at 0.3 to 1 ppm. Can cause eye and throat irritation. May cause the sensation of choking or burning throat or cause coughing, chest pain, difficult breathing, and erosion of the teeth. Other more serious symptoms that may develop within 24 hours are moderate to severe breathing difficulties and bluish skin color that may rapidly progress to death by swelling of the lung tissues or fluid in the lungs.

TLV/PEL	2 ppm	**IDLH**	25 ppm
STEL	4 ppm	**VP**	2.6 – 103 mm

Important Hazard Easily decomposes to form oxides of nitrogen (e.g., nitric oxide and nitrogen dioxide).

OXIDIZER — CORROSIVE — SPECIAL HAZARD — (Fuming) 4/1/OX — (> 40%) 4/0/OX — (< = 40%) 3/0

PRECAUTIONS

Avoid skin and eye contact. Use acid gloves, faceshield and safety glasses, apron and armguards. Use only within the confines of an air-exhausted hood.

FACESHIELD AND GLASSES — ACID GLOVES — APRON AND ARMGUARDS — AIR EXHAUSTED

Nitric Oxide

[10102–43–9]

NO
Nitrogen Monoxide

POISON

HAZARDS

Gas Sharp, sweet odor detectable at 0.3 to 1 ppm. Can be irritating to eyes, skin, and lungs. Prolonged exposure to high concentrations can cause lung damage.

TLV/PEL 25 ppm

IDLH 100 ppm

POISON   GAS HAZARD   OXIDIZER

PRECAUTIONS

Nitric oxide and nitrogen dioxide are formed when nitric acid is poured. Avoid eye exposure and inhalation. Use only within a closed, well-exhausted system.

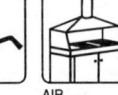

SAFETY GLASSES   AIR EXHAUSTED

Nitrogen Dioxide

[10102–44–0]

NO_2 **POISON**

HAZARDS

Gas Odor detectable at 0.06 to 0.2 ppm. 10 to 20 ppm is very irritating to eyes, nose and throat. High concentrations can cause severe lung damage.

TLV/PEL 3 ppm
STEL 5 ppm
IDLH 20 ppm

POISON   GAS HAZARD   OXIDIZER

PRECAUTIONS

Nitrogen dioxide and nitric oxide are formed when nitric acid is poured. Avoid eye exposure and inhalation. Use only within a closed, well-exhausted system.

SAFETY GLASSES   AIR EXHAUSTED

Nitrogen Trifluoride

[7783–54–2]

NF$_3$ **OXIDIZER**

HAZARDS

Gas Slightly moldy odor. Repeated exposure may cause liver and kidney damage. High concentrations can cause weakness, dizziness, and headache and damage blood cells. Prolonged exposure can cause spots on teeth and bone damage.

TLV/PEL 10 ppm

IDLH 1000 ppm

Important Hazard May be toxic at concentrations below the odor threshold. Do not use halon fire extinguishers on NF$_3$ fires.

OXIDIZER   GAS HAZARD   SPECIAL HAZARD   (Manufacturer)

PRECAUTIONS

Avoid all direct contact. Use only within a closed, well-exhausted system.

Note: This compound contributes to global warming if released to the atmosphere.

SAFETY GLASSES   AIR EXHAUSTED

Nitromethane

[75–52–5]

CH_3NO_2 **FLAMMABLE**

HAZARDS

Solid/Liquid Repeated contact can cause skin and eye irritation.

Vapor Mild, fruity odor detectable at 3.5 to 100 ppm. Can cause lung irritation and intoxication. High concentrations may cause liver and kidney damage. Flammable.

TLV/PEL	20 ppm	**FP**	95°F
IDLH	75 ppm	**AT**	785°F
VP	27.8 mm	**ER**	7.3% – 63%

FLAMMABLE

PRECAUTIONS

Avoid skin and eye contact. Use solvent gloves, faceshield and safety glasses, apron and armguards. Use only within the confines of an air-exhausted hood.
Keep away from ignition sources such as flame, spark, and heat.

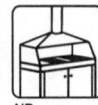

FACESHIELD AND GLASSES | SOLVENT GLOVES | APRON AND ARMGUARDS | AIR EXHAUSTED | NO IGNITION SOURCE

Nitrous Oxide

[10024–97–2]

N_2O
Nitrogen Oxide
Laughing Gas

OXIDIZER

HAZARDS

Gas Slightly sweet odor. Prolonged exposure may cause numbness, tingling, weakness, and blood damage.

TLV/PEL 50 ppm

Important Hazard Suspected of causing reproductive disorders.

GAS HAZARD | SPECIAL HAZARD | OXIDIZER

PRECAUTIONS

Use only within a closed, well-exhausted system. Avoid inhalation.

SAFETY GLASSES | AIR EXHAUSTED

Octane

[111–65–9]

C_8H_{18}

FLAMMABLE

HAZARDS

Liquid May cause eye irritation. Repeated contact may cause skin irritation.

Vapor Gasoline-like odor detectable at 15 to 235 ppm. May be irritating to eyes, nose, and throat. May also cause drowsiness, dizziness, confusion, nausea, and difficulty breathing. Flammable.

TLV/PEL	300 ppm
STEL	375 ppm
IDLH	1000 ppm
VP	11 mm
FP	56°F
AT	403°F
ER	1 – 6.5%

FLAMMABLE

PRECAUTIONS

Avoid skin and eye contact. Use solvent gloves, faceshield and safety glasses, apron and armguards. Use only within the confines of an air-exhausted hood.

Keep away from ignition sources such as flame, spark, and heat.

FACESHIELD AND GLASSES SOLVENT GLOVES APRON AND ARMGUARDS AIR EXHAUSTED NO IGNITION SOURCE

Octanol, Mixed Isomers [104–76–7]

$C_8H_{18}O$
Octyl Alcohol
2-Ethylhexanol

COMBUSTIBLE

HAZARDS

Liquid Can be readily absorbed through the skin. May cause eye irritation. Repeated contact may cause skin irritation.

Vapor/Mist Mild, slightly flower-like odor detectable at 0.001 to 2 ppm. High concentrations can cause eye, nose, and throat irritation. High concentrations can cause headache, dizziness, weakness, and drowsiness. Repeated exposures may cause liver and kidney damage. Combustible.

- **VP** 0.36 mm
- **FP** 140 – 178°F
- **AT** 448°F
- **ER** 0.88 – 9.7%

COMBUSTIBLE

PRECAUTIONS

Particular care should be taken to avoid skin and eye contact. Use solvent gloves, faceshield and safety glasses, apron and armguards. Use only within the confines of an air-exhausted hood.

Keep away from open flame and high temperatures.

FACESHIELD AND GLASSES SOLVENT GLOVES APRON AND ARMGUARDS AIR EXHAUSTED NO IGNITION SOURCE

Oxalic Acid

[144–62–7]

COOHCOOH

CORROSIVE (ACID)

HAZARDS

Dust/Mist Can cause chemical burns of the eyes, nose, throat, and skin. Strong poison when swallowed (displaces calcium in the body).

TLV/PEL   1 mg/m^3

IDLH   500 mg/m^3

STEL   2 mg/m^3

VP   < 0.001 mm

CORROSIVE

3 / 1 / 0

PRECAUTIONS

Avoid skin and eye contact. Use acid gloves, faceshield and safety glasses, apron and armguards.
Use in a well-ventilated area.

FACESHIELD AND GLASSES   ACID GLOVES   APRON AND ARMGUARDS   AIR EXHAUSTED

Oxygen

[7782-44-7]

O_2 **OXIDIZER**

HAZARDS

Liquid Freezes skin on contact.

Important Hazard Liquid oxygen and compressed oxygen can ignite organic materials spontaneously upon contact.

OXIDIZER   SPECIAL HAZARD

(Liquid)

PRECAUTIONS

Use cryogenic gloves when handling liquid. Keep liquid oxygen away from organic materials. Do not use grease on valves or fittings for compressed oxygen systems.

Use in a well-ventilated area.

SAFETY GLASSES   NO IGNITION SOURCE

Ozone

[10028–15–6]

O_3

POISON/OXIDIZER

HAZARDS

Gas Pungent odor detectable at 0.008 to 0.04 ppm. Exposure to 0.05 to 0.1 ppm for 13 to 30 minutes causes irritation and dryness of the throat. Above 0.1 ppm coughing, choking, chest pain, difficulty in breathing and blurred vision can occur. More severe exposures can also cause headache, dizziness, and a burning sensation in the eyes. Exposure to 0.6 to 0.8 ppm for 2 hours can result in difficulty in breathing for up to 24 hours.

TLV/PEL 0.05 ppm

STEL 0.2 ppm

IDLH 5 ppm

Important Hazard Toxic at low concentrations.

POISON OXIDIZER GAS HAZARD SPECIAL HAZARD

PRECAUTIONS

Use only within a closed, well-exhausted system. Avoid inhalation. Handle with care and keep away from heat.

SAFETY GLASSES AIR EXHAUSTED

Pentane

[109–66–0]

C_5H_{12}

FLAMMABLE

HAZARDS

Liquid May cause eye irritation. Repeated contact may cause skin irritation.

Vapor Gasoline-like odor detectable at 120 to 1150 ppm. No effects observed from exposure at 5000 ppm for 10 minutes. Higher concentrations may cause exhilaration, dizziness, and headache. Flammable.

TLV/PEL	600 ppm	**FP**	–57°F (n-pentane)
STEL	750 ppm	**AT**	500°F
IDLH	1500 ppm	**ER**	1.4 – 7.8%
VP	426 mm		

Important Hazard Vapors are extremely flammable.

FLAMMABLE   SPECIAL HAZARD

PRECAUTIONS

Avoid skin and eye contact. Use solvent gloves, faceshield and safety glasses, apron and armguards. Use only within the confines of an air-exhausted hood.

Keep away from ignition sources such as flame, spark, and heat.

FACESHIELD AND GLASSES   SOLVENT GLOVES   APRON AND ARMGUARDS   AIR EXHAUSTED   NO IGNITION SOURCE

Perchloric Acid

[7601–90–3]

HClO$_4$

OXIDIZER/ CORROSIVE (ACID)

HAZARDS

Liquid Can cause chemical burns to skin and eyes. Reacts violently with many organic and inorganic chemicals, including metals. Solutions of 50 to 72% are oxidizers.

Vapor Highly irritating to skin, eyes, nose, throat, and lungs. Prolonged exposure may cause severe coughing and vomiting.

Important Hazard High concentrations are unstable; can explode on shock or with heat.

OXIDIZER   CORROSIVE   SPECIAL HAZARD

(> 50%, < 72%)

PRECAUTIONS

Use only within the confines of an air-exhausted hood. Use acid gloves, faceshield and safety glasses, apron and armguards.

The precautions listed above apply to dilute concentrations of perchloric acid (i.e., 5% or less). If higher concentrations are needed, contact your safety department for additional precautions.

FACESHIELD AND GLASSES   ACID GLOVES   APRON AND ARMGUARDS   AIR EXHAUSTED

Periodic Acid

[10450–60–9]

$HIO_4 \cdot 2H_2O$

**OXIDIZER/
CORROSIVE (ACID)**

HAZARDS

Liquid/Dust Can cause chemical burns to skin and eyes. May be readily absorbed through the skin.

Dust/Mist Highly irritating to skin, eyes, and lungs. High concentrations may cause coughing, wheezing, headache, nausea, and vomiting.

OXIDIZER   CORROSIVE

PRECAUTIONS

Particular care should be taken to avoid skin and eye contact. Use acid gloves, faceshield and safety glasses, apron and armguards.
Use only within the confines of an air-exhausted hood.

FACESHIELD AND GLASSES   ACID GLOVES   APRON AND ARMGUARDS   AIR EXHAUSTED

Phenol

[108–95–2]

C_6H_5OH
Carbolic Acid

**POISON/CORROSIVE
(ACID)/COMBUSTIBLE**

HAZARDS

Solid/Liquid Can cause burns to skin and eyes. Can be readily absorbed through the skin. Skin absorption can cause headache, dizziness, muscular weakness, dimness of vision, ringing in the ears, irregular and rapid breathing, and weak pulse.

Vapor Medicinal odor detectable at 0.005 to 1 ppm. Can cause irritation of eyes, nose, and lungs. High concentrations can cause damage to lungs, heart, liver, and kidneys. Combustible.

TLV/PEL 5 ppm   **VP** 0.36 mm   **AT** 1319°F
IDLH 250 ppm   **FP** 174°F   **ER** 1.7 – 8.6%

Important Hazard Liquid can be absorbed through the skin in amounts that can cause death.

POISON   CORROSIVE   COMBUSTIBLE   SPECIAL HAZARD

4 / 2 / 0

PRECAUTIONS

Avoid all skin and eye contact. Use solvent gloves, faceshield and safety glasses, apron and armguards. Use only within the confines of an air-exhausted hood.

Keep away from flame and high temperatures.

Particular care should be taken to avoid skin contact with phenol. May require the use of gloves specifically designed for protection against phenol.

FACESHIELD AND GLASSES   SOLVENT GLOVES   APRON AND ARMGUARDS   AIR EXHAUSTED   NO IGNITION SOURCE

Phenylethylamine

[64–04–0]

C$_6$H$_5$(CH$_2$)$_2$NH$_2$
PEA
Phenethylamine

**CORROSIVE (BASE)/
COMBUSTIBLE**

HAZARDS

Liquid Can be irritating to skin and eyes. May cause an allergic reaction in some susceptible individuals.

Vapor Can cause irritation of skin and eyes. May cause an allergic reaction in some susceptible individuals. Combustible.

FP 195°F

CORROSIVE   COMBUSTIBLE

PRECAUTIONS

Avoid skin and eye contact. Use solvent gloves, faceshield and safety glasses, apron and armguards. Use only within the confines of an air-exhausted hood.

FACESHIELD AND GLASSES   SOLVENT GLOVES   APRON AND ARMGUARDS   AIR EXHAUSTED

Phenyltrichlorosilane

[98–13–5]

$C_6H_5SiCl_3$
Trichlorophenylsilane

CORROSIVE (ACID)/ COMBUSTIBLE

HAZARDS

Liquid Can be irritating to skin and eyes.

Vapor Strong irritating odor detectable at 10 ppm. Can be irritating to eyes, nose, throat, and lungs. Combustible.

VP 0.5 mm

FP 196°F

Important Hazard Will react with water to form hydrochloric acid (HCl).

CORROSIVE COMBUSTIBLE SPECIAL HAZARD

PRECAUTIONS

Avoid inhalation and skin and eye contact. Use acid gloves, faceshield and safety, apron and armguards. Use only within the confines of an air-exhausted hood.
Keep away from open flame and high temperatures.
Keep away from water, including small amounts of moisture.

FACESHIELD AND GLASSES   ACID GLOVES   APRON AND ARMGUARDS   AIR EXHAUSTED   NO IGNITION SOURCE   WATER REACTIVE

Phosphine

[7803–51–2]

PH$_3$

POISON/FLAMMABLE

HAZARDS

Gas Fishy or garlic-like odor detectable at 0.01 to 5 ppm. Can cause eye, nose, and lung irritation. Exposures averaging under 10 ppm can cause headache, breathing difficulties, chest tightness, cough, loss of appetite, abdominal pain, giddiness, numbness, lethargy, nausea, vomiting, and diarrhea. Flammable.

TLV/PEL	0.3 ppm	**AT**	100°F
STEL	1 ppm	**ER**	1.6% – ?
IDLH	50 ppm		

Important Hazard Poisonous at low concentrations. High concentrations (> 15%) are pyrophoric — ignite and burn on contact with air.

POISON   FLAMMABLE   GAS HAZARD   SPECIAL HAZARD   AIR REACTIVE

PRECAUTIONS

Use only within a closed, well-exhausted system. Use only with a toxic gas monitor present and operating.

Fire extinguishers are ineffective in controlling fires from gas leaks; the gas supply must be turned off. Report all suspected exposures to your medical department immediately.

Keep away from ignition sources such as flame, spark, and heat.

SAFETY GLASSES   AIR EXHAUSTED   NO IGNITION SOURCE

Phosphoric Acid

[7664–38–2]

H_3PO_4

CORROSIVE (ACID)

HAZARDS

Liquid   Odorless liquid. Can cause severe burns to skin and eyes.

Mist   Low concentrations can be irritating to eyes, nose, and throat. Inhalation of 3.6 to 11.3 mg/m^3 can cause coughing. High concentrations can cause chemical burns to skin, eyes, and lungs. Long-term exposure to high concentrations can cause erosion of teeth.

TLV/PEL   1 mg/m^3
IDLH   1000 mg/m^3
STEL   3 mg/m^3
VP   0.03 mm

CORROSIVE

PRECAUTIONS

Avoid skin and eye contact. Use acid gloves, faceshield and safety glasses, apron and armguards. Use only within the confines of an air-exhausted hood.

FACESHIELD AND GLASSES   ACID GLOVES   APRON AND ARMGUARDS   AIR EXHAUSTED

Phosphorus, Red

[7723–14–0]

P_4

COMBUSTIBLE

Phosphorus, Amorphous

HAZARDS

Dust   Can be irritating to skin, eyes, and lungs. Flammable.

IDLH   5 mg/m^3

AT   500°F

Important Hazard   May ignite with friction or on contact with oxidizers. Reacts with strong alkali to form phosphine gas.

COMBUSTIBLE   SPECIAL HAZARD

PRECAUTIONS

Avoid skin and eye contact. Use acid or solvent gloves and safety glasses. Use in a well-ventilated area.
Keep away from open flame and high temperatures.

ACID GLOVES   SAFETY GLASSES   NO IGNITION SOURCE

Phosphorus Oxychloride [10025–87–3]

POCl₃
POCL

CORROSIVE (ACID)

HAZARDS

Liquid Can cause burns to skin and eyes.

Vapor Odor detectable at low concentrations. Can be irritating to skin, eyes, and lungs. May cause dizziness, headache, weakness, loss of appetite, nausea, cough, and lung damage.

TLV/PEL 0.1 ppm

VP 40 mm at 27.3°C

Important Hazard Will react with water to form hydrochloric acid, phosphorus pentoxide, and phosphoric acid (fumes).

CORROSIVE   SPECIAL HAZARD

4 / 0 / 2 / W

PRECAUTIONS

Avoid skin and eye contact. Use acid gloves, faceshield and safety glasses, apron and armguards. Use only within the confines of a closed, well-exhausted system.

Keep away from water, including small amounts of moisture.

FACESHIELD AND GLASSES   ACID GLOVES   APRON AND ARMGUARDS   AIR EXHAUSTED   WATER REACTIVE

322

Phosphorus Pentafluoride [7647–19–0]

PF$_5$

CORROSIVE (ACID)

HAZARDS

Gas Can be irritating to skin and eyes. High concentrations can damage nose, throat, and lungs, and may cause fluid in lungs.

Important Hazard Forms hydrogen fluoride (HF) gas on contact with water or moisture in the air. Skin and eye contact is not immediately evident; pain may not start for up to 24 hours.

CORROSIVE   GAS HAZARD   SPECIAL HAZARD

PRECAUTIONS

Avoid skin and eye exposure and inhalation. Use only within a closed system.

Keep away from water, including small amounts of moisture.

SAFETY GLASSES   AIR EXHAUSTED   WATER REACTIVE

Phosphorus Pentoxide [1314–56–3]

P_2O_5
Phosphoric Anhydride

CORROSIVE (ACID)

HAZARDS

Solid Can be irritating to skin and eyes.

Dust Exposure to 3.6 to 11.3 mg/m^3 can cause coughing. May be irritating to skin, eyes, and lungs.

Important Hazard Reacts with water to evolve heat and phosphoric acid.

CORROSIVE   SPECIAL HAZARD

(Manufacturer)

PRECAUTIONS

Phosphorus pentoxide is formed when phosphorus oxychloride decomposes. Avoid skin and eye contact. Use acid gloves, faceshield and safety glasses, apron and armguards. Use in a well-ventilated area.
Keep away from water, including small amounts of moisture.

FACESHIELD AND GLASSES   ACID GLOVES   APRON AND ARMGUARDS   WATER REACTIVE

Phosphorus Tribromide [7789–60–8]

PBr$_3$
Phosphorus Bromide

**CORROSIVE (ACID)/
POISON**

HAZARDS

Liquid Can cause burns to skin and eyes.

Vapor Penetrating odor detectable at low concentrations. Can be irritating to skin, eyes, and lungs. Higher concentrations can cause fluid in the lungs and death. Symptoms can be delayed for 2 to 24 hours.

VP 10 mm at 47.8°C

Important Hazard Will react with water to form spontaneously flammable phosphine gas along with hydrobromic acid and phosphorous acids, which in turn can react with most metals to generate hydrogen, a flammable gas.

POISON    CORROSIVE    SPECIAL HAZARD

PRECAUTIONS

Avoid skin and eye contact. Use acid gloves, faceshield and safety glasses, apron and armguards. Use only within the confines of a closed, well-exhausted system.

Keep away from water, including small amounts of moisture.

FACESHIELD AND GLASSES    ACID GLOVES    APRON AND ARMGUARDS 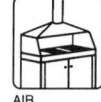   AIR EXHAUSTED    WATER REACTIVE

α-Pinene

[80–56–8]

$C_{10}H_{16}$ **FLAMMABLE**

HAZARDS

Liquid Can cause skin and eye irritation. May be readily absorbed through the skin.

Vapor Pleasant, pine-like odor detectable at 0.003 to 0.06 ppm. High concentrations may cause throat and lung irritation, mental disturbance, difficulty walking, and kidney damage. Repeated exposure to high concentrations can cause rapid heartbeat, dizziness, nervous disturbances, chest pain, bronchitis, and kidney damage. Flammable.

VP 5 mm @ 25°C

FP 91°F

AT 491°F

FLAMMABLE

PRECAUTIONS

Particular care should be taken to avoid skin and eye contact. Use solvent gloves, faceshield and safety glasses, apron and armguards. Use only within the confines of an air-exhausted hood.

Keep away from ignition sources such as flame, spark, and heat.

FACESHIELD AND GLASSES | SOLVENT GLOVES | APRON AND ARMGUARDS | AIR EXHAUSTED | NO IGNITION SOURCE

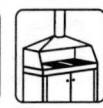

Platinum

[7440–06–4]

Pt

UNCLASSIFIED

HAZARDS

Solid/Liquid Platinum salts can cause skin rash and an allergic skin reaction.

Dust/Mist Can cause nose, throat, and lung irritation. Can cause allergic inflammation of the nose. Platinum salts can cause asthma attacks.

TLV/PEL 1.0 mg/m^3 (metal dust)
0.002 mg/m^3 (soluble salts, as platinum)

(Manufacturer)

PRECAUTIONS

Avoid skin and eye contact. Use acid or solvent gloves, faceshield and safety glasses, apron and armguards.

Platinum salts should be used only within the confines of an air-exhausted hood.

Platinum metals should be used in a well-ventilated area.

FACESHIELD AND GLASSES   ACID GLOVES   APRON AND ARMGUARDS   AIR EXHAUSTED

Polyethylene Glycol [25322–68–3]

Carbowax

COMBUSTIBLE

HAZARDS

Liquid/Solid Practically nontoxic.

Vapor Practically nontoxic. Vapors generated at high temperatures may cause eye, nose, and throat irritation. Combustible.

VP < 0.01 mm

FP 300°F

Important Hazard Polyethylene glycol contains trace amounts of ethylene oxide, which is a suspected carcinogen. Accumulations of ethylene oxide may occur in the headspace of shipping and storage containers or in enclosed areas where the product is being handled or used.

COMBUSTIBLE   SPECIAL HAZARD   CANCER CAUSING

PRECAUTIONS

Avoid skin and eye contact. Use solvent gloves and safety glasses. Use in a well-ventilated area.
Keep away from open flame and high temperatures.

SOLVENT GLOVES   SAFETY GLASSES   NO IGNITION SOURCE

Potassium Bromide

[7758–02–3]

KBr

UNCLASSIFIED

HAZARDS

Dust/Mist May be irritating to skin, eyes, and lungs. Prolonged exposure may cause acne. Exposure to high concentrations may cause drowsiness.

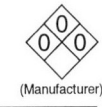

(Manufacturer)

PRECAUTIONS

Avoid skin and eye contact. Use acid or solvent gloves and safety glasses. Use in a well-ventilated area.

ACID GLOVES   SAFETY GLASSES

Potassium Carbonate [584–08–7]

K_2CO_3 **UNCLASSIFIED**

HAZARDS

Dust/Mist May be irritating to skin, eyes, nose, and throat. May cause coughing, sneezing, and difficulty breathing.

(Manufacturer)

PRECAUTIONS

Avoid skin and eye contact. Use acid or solvent gloves and safety glasses. Use in a well-ventilated area.

ACID GLOVES   SAFETY GLASSES

Potassium Chloride [7447-40-7]

KCl

UNCLASSIFIED

HAZARDS

Dust/Mist Can be mildly irritating to skin, eyes, nose, and throat. High concentrations may cause nausea, blood changes, and irregular heartbeat.

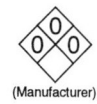

(Manufacturer)

PRECAUTIONS

Avoid skin and eye contact. Use acid or solvent gloves and safety glasses. Use in a well-ventilated area.

ACID GLOVES   SAFETY GLASSES

Potassium Chromate

[7789–00–6]

K$_2$CrO$_4$ **OXIDIZER**
Bipotassium Chromate
Dipotassium Chromate

HAZARDS

Dust/Mist Can be highly irritating to skin, eyes, nose, and throat. Can cause asthmatic bronchitis and liver and kidney damage. Can cause lung cancer.

TLV/PEL 0.05 mg/m^3 (as hexavalent chromium)

Important Hazard Confirmed human carcinogen.

OXIDIZER

CANCER CAUSING

1 / 3 / 0 (Manufacturer)

PRECAUTIONS

Avoid skin and eye contact. Use acid gloves and safety glasses. Use within the confines of an air-exhausted hood.

ACID GLOVES

SAFETY GLASSES

AIR EXHAUSTED

Potassium Cyanate

[590–28–3]

KOCN

UNCLASSIFIED

HAZARDS

Solid/Liquid May cause skin and eye irritation. May be slowly transformed in the body to cyanide, but does not have the high toxicity of cyanides.

(Manufacturer)

PRECAUTIONS

Avoid skin and eye contact. Use solvent or acid gloves and safety glasses. Use in a well-ventilated area.

SAFETY GLASSES

SOLVENT GLOVES

Potassium Dichromate

[7778–50–9]

K$_2$Cr$_2$O$_7$

OXIDIZER

Potassium Bichromate

HAZARDS

Dust/Mist Can be highly irritating to skin, eyes, nose, and throat. Can cause asthmatic bronchitis and liver and kidney damage. Can cause lung cancer.

TLV/PEL 0.05 mg/m^3 (as hexavalent chromium)

Important Hazard Confirmed human carcinogen.

OXIDIZER

CANCER CAUSING

(Manufacturer)

PRECAUTIONS

Avoid skin and eye contact. Use acid gloves and safety glasses. Use within the confines of an air-exhausted hood.

ACID GLOVES

SAFETY GLASSES

AIR EXHAUSTED

Potassium Hydrogen Phthalate

[877–24–7]

KHC$_8$H$_4$O$_4$
Potassium Acid Phthalate
Potassium Biphthalate

**UNCLASSIFIED
(WEAK ACID)**

HAZARDS

Dust/Mist May be irritating to skin, eyes, and lungs. May cause cough and sore throat.

PRECAUTIONS

Avoid skin and eye contact. Use solvent gloves and safety glasses.
Use in a well-ventilated area.

SOLVENT GLOVES

SAFETY GLASSES

Potassium Hydroxide [1310–58–3]

KOH
Lye

CORROSIVE (BASE)

HAZARDS

Solid/Liquid Can cause severe burns to skin and eyes. Contact with dilute solution can cause skin rash or small sores on exposed skin.

Dust/Mist Can be highly irritating to skin, eyes, nose, throat, and lungs. However, severe injury is usually avoided because sneezing, coughing, and discomfort make high exposure intolerable.

TLV/PEL 2 mg/m^3 C

CORROSIVE

PRECAUTIONS

Avoid skin and eye contact. Use acid gloves, faceshield and safety glasses, apron and armguards.
Use in a well-ventilated area.

FACESHIELD AND GLASSES   ACID GLOVES   APRON AND ARMGUARDS

Potassium Iodate

[7758–05–6]

KIO_3

OXIDIZER

HAZARDS

Dust/Mist May be irritating to eyes, nose, throat, and lungs. Prolonged exposure to the skin may cause irritation. May cause kidney and blood cell damage.

OXIDIZER

(Manufacturer)

PRECAUTIONS

Avoid skin and eye contact. Use acid or solvent gloves and safety glasses. Use in a well-ventilated area.

ACID GLOVES   SAFETY GLASSES

Potassium Iodide

[7681-11-0]

KI

UNCLASSIFIED

HAZARDS

Solid/Liquid May cause skin and eye irritation.

Dust/Mist May be irritating to eyes, nose, and throat, and cause coughing and chest discomfort. Repeated exposure may cause skin rash, running nose, headache and irritation; weakness, anemia, weight loss, and general depression may also occur.

2 / 0 / 1 (Manufacturer)

PRECAUTIONS

Avoid skin and eye contact. Use solvent or acid gloves and safety glasses. Use in a well-ventilated area.

SAFETY GLASSES   SOLVENT GLOVES

Potassium Nitrate

[7757–79–1]

KNO_3

OXIDIZER

HAZARDS

Dust/Mist May be irritating to eyes, nose, throat, and lungs. Prolonged exposure to the skin may cause irritation. May cause kidney and blood cell damage.

Important Hazard When heated to decomposition (about 752°F) emits poisonous oxides of nitrogen and potassium oxide.

OXIDIZER | SPECIAL HAZARD | (Manufacturer)

PRECAUTIONS

Avoid skin and eye contact. Use acid or solvent gloves and safety glasses. Use in a well-ventilated area.

ACID GLOVES | SAFETY GLASSES

Potassium Permanganate [7722-64-7]

KMnO$_4$ **OXIDIZER**

HAZARDS

Dust/Mist High concentrations may cause irritation to skin, eyes, nose, throat, and lungs with sore throat, coughing, shortness of breath, and difficult breathing.

TLV/PEL 0.2 mg/m^3 (as Mn)

OXIDIZER

(Manufacturer)

PRECAUTIONS

Avoid skin and eye contact. Use acid gloves and safety glasses.
Use in a well-ventilated area.

ACID GLOVES   SAFETY GLASSES

Potassium Phosphate

[7778-53-2]

K_3PO_4

UNCLASSIFIED

HAZARDS

Dust/Mist Can be mildly irritating to skin, eyes, nose, and throat.

PRECAUTIONS

Avoid skin and eye contact. Use acid or solvent gloves and safety glasses. Use in a well-ventilated area.

ACID GLOVES   SAFETY GLASSES

Potassium Sulfate

[7778–80–5]

K_3SO_4

UNCLASSIFIED

HAZARDS

Dust/Mist Can be mildly irritating to skin, eyes, and lungs.

(Manufacturer)

PRECAUTIONS

Avoid skin and eye contact. Use acid or solvent gloves and safety glasses. Use in a well-ventilated area.

ACID GLOVES SAFETY GLASSES

Propane

[74–98–6]

C_3H_8

FLAMMABLE

HAZARDS

Gas Faint odor detectable at 12,000 to 20,000 ppm when pure (fuel grades have mercaptan odorant added that gives the propane mixture a skunk-like odor detectable at low concentrations). Can displace oxygen in air (acts as a simple asphyxiant). 100,000 ppm can cause slight dizziness after a few minutes of exposure. Flammable.

TLV/PEL	1000 ppm
IDLH	2100 ppm
AT	842°F
ER	2.2 – 9.5%

FLAMMABLE

PRECAUTIONS

Use in a well-ventilated area.
Keep away from all ignition sources such as flame, spark, and heat.

NO IGNITION SOURCE

Propionic Acid

[79–09–4]

CH$_3$CH$_2$COOH
Methylacetic Acid

**CORROSIVE (ACID)/
COMBUSTIBLE**

HAZARDS

Liquid Can cause severe burns to skin and eyes.

Vapor Unpleasant odor detectable at 0.03 to 0.2 ppm. Can be irritating to skin and eyes. High concentrations can cause cough and an asthmatic response. Combustible.

TLV/PEL	10 ppm
VP	2.9 mm
FP	130°F
AT	905°F
ER	2.9 – 15%

CORROSIVE COMBUSTIBLE

PRECAUTIONS

Avoid skin and eye contact. Use acid gloves, faceshield and safety glasses, apron and armguards. Use only within the confines of an air-exhausted hood. Keep away from open flame and high temperatures.

FACESHIELD AND GLASSES   ACID GLOVES   APRON AND ARMGUARDS   AIR EXHAUSTED   NO IGNITION SOURCE

1,2-Propylene Glycol

[57–55–6]

COMBUSTIBLE

CH$_3$CHOHCH$_2$OH
1,2-Propanediol
Methylene Glycol

HAZARDS

Liquid Repeated contact may cause skin and eye irritation.

Vapor Practically nontoxic. Vapors generated when liquid is heated may be irritating to eyes and lungs. Combustible. This chemical is not likely to form significant vapor concentrations at room temperature.

- **VP** 0.08 mm
- **FP** 210°F
- **AT** 700°F
- **ER** 2.6 – 12.6%

COMBUSTIBLE

PRECAUTIONS

Avoid skin and eye contact. Use solvent gloves and safety glasses.
Use in a well-ventilated area. If heated, use in the confines of an air-exhausted hood.
Keep away from open flame and high temperatures.

SAFETY GLASSES   SOLVENT GLOVES   AIR EXHAUSTED

Propylene Glycol Monomethyl Ether

[107–98–2]

CH₃OCH₂CHOHCH₃
1-Methoxy-2-Propanol
PGME

FLAMMABLE

HAZARDS

Liquid Repeated contact may cause skin and eye irritation.

Vapor Sweet, ether-like odor detectable at about 10 ppm. Exposure at 300 to 400 ppm can cause light-headedness and mild eye irritation. 500 ppm can cause eye, nose, and throat irritation. At 700 ppm, tearing and runny nose can occur. Above 1000 ppm headache, nausea, light-headedness, drowsiness, incoordination, or possible unconsciousness can occur, but these concentrations are intolerable because of severe eye, nose, and throat irritation. Flammable.

TLV/PEL	100 ppm	**FP**	90°F
STEL	150 ppm	**AT**	547°F
VP	11.8 mm @ 25°C	**ER**	1.6 – 13.8%

FLAMMABLE

PRECAUTIONS

Avoid skin and eye contact. Use solvent gloves, faceshield and safety glasses, apron and armguards. Use only within the confines of an air-exhausted hood.
Keep away from ignition sources such as flame, spark, and heat.

FACESHIELD AND GLASSES | SOLVENT GLOVES | APRON AND ARMGUARDS | AIR EXHAUSTED | NO IGNITION SOURCE

Propylene Glycol Monomethyl Ether Acetate [108–65–6]

PM Acetate
1-Methoxy-2-Propyl Acetate
PGMEA

COMBUSTIBLE

HAZARDS

Liquid Repeated contact may cause skin and eye irritation.

Vapor Sweet ether-like odor detectable at low concentrations. Effects similar to propylene glycol monomethyl ether are expected but may be more irritating. May cause eye, nose, and throat irritation. High concentrations may cause dizziness. Combustible.

VP 3.7 mm

FP 108°F

AT 522°F

ER 1.3 – 13.1%

PRECAUTIONS

Avoid skin and eye contact. Use solvent gloves, faceshield and safety glasses, apron and armguards. Use only within the confines of an air-exhausted hood.
Keep away from open flame and high temperatures.

FACESHIELD AND GLASSES | SOLVENT GLOVES | APRON AND ARMGUARDS | AIR EXHAUSTED | NO IGNITION SOURCE

347

Pyridine

[110–86–1]

C_5H_5N
Azine
Azabenzene

FLAMMABLE

HAZARDS

Liquid Can be readily absorbed through the skin. Repeated contact can cause skin and eye irritation.

Vapor Unpleasant odor detectable at 0.02 to 2 ppm. Can cause eye and nose irritation. Repeated exposure at 6 to 12 ppm can cause headache, dizziness, sleeplessness, nausea, and vomiting. Repeated exposure to higher concentrations may cause liver and kidney damage. Flammable.

TLV/PEL	5 ppm	**FP**	68°F
IDLH	1000 ppm	**AT**	900°F
VP	18 mm	**ER**	1.8 – 12.4%

FLAMMABLE

PRECAUTIONS

Particular care should be taken to avoid skin and eye contact. Use solvent gloves, faceshield and safety glasses, apron and armguards. Use only within the confines of an air-exhausted hood.

Keep away from ignition sources such as flame, spark, and heat.

FACESHIELD AND GLASSES   SOLVENT GLOVES   APRON AND ARMGUARDS   AIR EXHAUSTED   NO IGNITION SOURCE

Pyrogallol

[87–66–1]

$C_6H_3(OH)_3$
Pyrogallic Acid

CORROSIVE (ACID)

HAZARDS

Solid/Liquid Can cause skin and eye irritation. Can be readily absorbed through the skin in amounts that may be poisonous.

Vapor/Mist/Dust May cause irritation of eyes, nose, and throat. High concentrations may cause convulsions, damage to blood cells, kidney and liver damage.

Important Hazard Repeated contact with the skin can cause an allergic reaction to the chemical.

POISON   CORROSIVE   SPECIAL HAZARD

PRECAUTIONS

Avoid skin and eye contact. Use solvent gloves, faceshield and safety glasses, apron and armguards. Use only within the confines of an air-exhausted hood.

Particular care should be taken to avoid skin contact with pyrogallol.

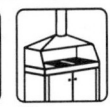

FACESHIELD AND GLASSES   ACID GLOVES   APRON AND ARMGUARDS   AIR EXHAUSTED

Refractory Ceramic Fiber [65997–13–3]

Kaowool

UNCLASSIFIED

HAZARDS

Fiber Can be irritating to skin, eyes, and lungs. May cause lung damage.

Important Hazard Suspected carcinogen.

CANCER CAUSING

PRECAUTIONS

Use acid or solvent gloves and safety glasses. Surfaces contaminated with the fibers should be cleaned by dampening the area and wiping with damp towels. Dry sweeping or brushing must be avoided.

Use in a well-ventilated area or within the confines of an air-exhausted hood.

ACID GLOVES   SAFETY GLASSES   AIR EXHAUSTED

Resins

Epoxy Resins
Novolac Resins
Phenolic Resins
Polyimide Resins

UNCLASSIFIED

HAZARDS

Dust/Mist May cause skin, eye, and lung irritation, headaches, and nausea. Many resins are flammable or combustible.

Important Hazard Contact the safety department or refer to MSDS for hazard information on specific compounds. Some may cause allergic skin reactions.

SPECIAL HAZARD

(as Epoxy Resin, Manufacturer)

PRECAUTIONS

Avoid skin and eye contact. Use solvent gloves and safety glasses. Use in a well-ventilated area.

SOLVENT GLOVES

SAFETY GLASSES

351

Selenium

[7782–49–2]

Se **POISON**

HAZARDS

Solid Can cause skin and eye irritation.

Dust Can cause irritation of skin, eyes, nose, and throat. Repeated exposure may cause stomach upset, dizziness, giddiness, depression, weakness, and garlic odor of the breath and sweat. Higher concentrations can cause loss of fingernails and hair. May cause damage to liver, kidneys, and blood.

TLV/PEL 0.2 mg/m^3

IDLH 1 mg/m^3

POISON

PRECAUTIONS

Avoid skin and eye contact. Use acid or solvent gloves and safety glasses. Use only within the confines of an air-exhausted hood.

ACID GLOVES   SAFETY GLASSES   AIR EXHAUSTED

Silane

[7803-62-5]

FLAMMABLE

SiH_4
Silicon Tetrahydride

HAZARDS

Gas Repulsive odor detectable at 0.5 ppm. Unreacted silane in air may cause headache and nausea.

TLV/PEL 5 ppm

ER 1 – 96%

Important Hazard Greater than 3% silane concentrations are pyrophoric — ignites and burns on contact with air. High concentrations of silane are not always pyrophoric immediately, but may explode if the gas release continues.

FLAMMABLE   GAS HAZARD   AIR REACTIVE   SPECIAL HAZARD

4/1/3

PRECAUTIONS

Use only within a closed, well-exhausted system.
Keep away from ignition sources such as flame, spark, and heat.
Fire extinguishers are ineffective in controlling fires from gas leaks; the gas supply must be turned off.

SAFETY GLASSES   AIR EXHAUSTED   NO IGNITION SOURCE

Silica, Amorphous

[7631–86–9]

SiO$_2$
Silicon Dioxide

UNCLASSIFIED

HAZARDS

Dust Can be mildly irritating to eyes and lungs. Respirable (dust fine enough to get into lungs) fume silica and fused silica can accumulate in lungs and cause lung damage.

TLV/PEL 10 mg/m^3 (diatomaceous earth; precipitated silica; silica gel)
2 mg/m^3 (respirable fume silica)

IDLH 0.1 mg/m^3 (respirable fused silica)
3000 mg/m^3

(Manufacturer)

PRECAUTIONS

Amorphous silica is formed as a breakdown product of silane, dichlorosilane, and silicon tetrachloride. It is the major component of the white residue formed when silane burns.

Avoid skin and eye contact. Use acid or solvent gloves and safety glasses. Use in a well-ventilated area.

ACID GLOVES   SAFETY GLASSES

Silicon

[7440–21–3]

Si

UNCLASSIFIED

HAZARDS

Dust Can be mildly irritating to eyes and lungs. Minimum explosive concentration 160 g/m^3 (16%).

TLV/PEL 10 mg/m^3

Important Hazard A dust cloud of this material can explode under certain circumstances.

SPECIAL HAZARD

PRECAUTIONS

Avoid skin and eye contact. Use acid or solvent gloves and safety glasses.
Use in a well-ventilated area.
Keep dust away from ignition sources such as flame, spark, and heat.

ACID GLOVES   SAFETY GLASSES   NO IGNITION SOURCE

Silicone

Polysiloxanes
RTV Silicone Rubber

COMBUSTIBLE

HAZARDS

Paste Uncured material has a vinegar-like odor. Can be irritating to eyes. Uncured material may be combustible.

VP < 5 mm

FP 210°F (for some, most are higher)

COMBUSTIBLE

(Manufacturer)

PRECAUTIONS

Avoid eye contact. Use safety glasses. Use in a well-ventilated area. Keep uncured material away from flame and high temperatures.

SAFETY GLASSES

NO IGNITION SOURCE

Silicon Tetrachloride

[10026–04–7]

CORROSIVE (ACID)

SiCl$_4$
Siltet
Silicon Chloride

HAZARDS

Liquid Can cause burns to skin and eyes.

Gas/Vapor Acid odor detectable at low concentrations. Can be irritating to skin, eyes, and lungs.

Important Hazard Reacts with water and moisture in the air to form hydrochloric acid.

CORROSIVE   GAS HAZARD   SPECIAL HAZARD

PRECAUTIONS

Avoid skin and eye contact. Use acid gloves, faceshield and safety glasses, apron and armguards. Use only within the confines of an air-exhausted hood or closed system.

Keep away from water, including small amounts of moisture.

FACESHIELD AND GLASSES   ACID GLOVES   APRON AND ARMGUARDS   AIR EXHAUSTED   WATER REACTIVE

Silicon Tetrafluoride

[7783–61–1]

SiF$_4$
Tetrafluorosilane

CORROSIVE (ACID)

HAZARDS

Gas Sharp, irritating, and suffocating odor detectable at low concentrations. Can cause irritation of the nose, eyes, lungs and skin. Higher exposures can cause severe irritation of the eyes and eyelids, lung damage, and circulatory collapse. 50 ppm may be fatal if inhaled for 30 to 60 minutes.

Important Hazard Reacts with water and moisture in the air to form hydrofluoric acid.

CORROSIVE   GAS HAZARD   SPECIAL HAZARD

0 / 3 / 2 / W

PRECAUTIONS

Use only within a closed, well-exhausted system. Avoid all direct contact. Keep away from water, including small amounts of moisture.

SAFETY GLASSES   AIR EXHAUSTED   WATER REACTIVE

Silver and Compounds

[7440–22–4]

Ag

UNCLASSIFIED

HAZARDS

Dust/Fume Can cause a blue-gray discoloration of the skin, eyes, nose, throat, and mouth.

TLV/PEL 0.1 mg/m^3 (metal dust and fume)
0.01 mg/m^3 (soluble compounds as silver)

IDLH 10 mg/m^3

(Manufacturer)

PRECAUTIONS

Avoid skin and eye contact. Use acid or solvent gloves and safety glasses. Use only within the confines of an air-exhausted hood.

ACID GLOVES   SAFETY GLASSES   AIR EXHAUSTED

359

Silver Acetate

[563–63–3]

CH₃COOAg **COMBUSTIBLE**

HAZARDS

Dust Can cause a blue-gray discoloration of the skin, eyes, nose, throat, and mouth. High concentrations may cause headache, coughing, dizziness, and difficulty in breathing.

TLV/PEL 0.01 mg/m^3 (soluble compounds as silver)

IDLH 10 mg/m^3 (as silver)

PRECAUTIONS

Avoid skin and eye contact. Use acid or solvent gloves and safety glasses. Use only within the confines of an air-exhausted hood.

ACID GLOVES   SAFETY GLASSES   AIR EXHAUSTED

Silver Nitrate

[7761–88–8]

AgNO$_3$

OXIDIZER

HAZARDS

Dust Can cause a blue-gray discoloration of the skin, eyes, nose, throat, and mouth. Can irritate nose, throat, and lungs. May cause sore throat, coughing, and shortness of breath.

TLV/PEL 0.01 mg/m^3 (soluble compounds as silver)

IDLH 10 mg/m^3 (as silver)

OXIDIZER

(Manufacturer)

PRECAUTIONS

Avoid skin and eye contact. Use acid or solvent gloves and safety glasses. Use only within the confines of an air-exhausted hood.

ACID GLOVES   SAFETY GLASSES   AIR EXHAUSTED

Sodium Acetate

[127–09–3]

$NaC_2H_3O_2$

COMBUSTIBLE

HAZARDS

Dust/Mist Repeated contact may cause skin irritation. May cause eye, nose, and throat irritation. Combustible.

AT 1125°F

COMBUSTIBLE

PRECAUTIONS

Avoid skin and eye contact. Use solvent or acid gloves and safety glasses. Use in a well-ventilated area. Keep away from open flame and high temperatures.

SAFETY GLASSES   SOLVENT GLOVES   NO IGNITION SOURCE

Sodium Bicarbonate

[144–55–8]

NaHCO$_3$
Baking Soda

UNCLASSIFIED

HAZARDS

Dust/Mist Repeated contact may cause skin irritation. May cause eye, nose, and throat irritation.

TLV/PEL 10 mg/m^3 (as a nuisance dust)

(Manufacturer)

PRECAUTIONS

Avoid skin and eye contact. Use solvent or acid gloves and safety glasses. Use in a well-ventilated area.

SAFETY GLASSES

SOLVENT GLOVES

Sodium Bisulfate

[7681-38-1]

NaHSO$_4$

ACID

HAZARDS

Solid/Liquid Can be irritating to skin and eyes.

Dust/Mist Can be irritating to eyes, skin, nose, and throat. Exposure to higher concentrations can cause headache and choking.

Important Hazard Emits very toxic fumes of sodium monoxide and oxides of sulfur when heated to decomposition.

CORROSIVE SPECIAL HAZARD

1 / 0 / 1 (Manufacturer)

PRECAUTIONS

Avoid skin and eye contact. Use acid gloves and safety glasses.
Use in a well-ventilated area.

ACID GLOVES SAFETY GLASSES

Sodium Bisulfite

[7631–90–5]

NaHSO$_3$

UNCLASSIFIED

HAZARDS

Solid Can be irritating to skin and eyes.

Dust Can be irritating to eyes, skin, nose, and throat.

TLV/PEL 5 mg/m^3

Important Hazard Emits very toxic fumes of sodium monoxide and oxides of sulfur when heated to decomposition.

SPECIAL HAZARD

PRECAUTIONS

Avoid skin and eye contact. Use acid gloves and safety glasses.
Use in a well-ventilated area.

ACID GLOVES   SAFETY GLASSES

Sodium Carbonate

[497–19–8]

Na$_2$CO$_3$
Soda Ash

UNCLASSIFIED

HAZARDS

Dust Repeated contact may cause skin irritation. May cause eye, nose, throat, and lung irritation.

(Manufacturer)

PRECAUTIONS

Avoid skin and eye contact. Use solvent or acid gloves and safety glasses. Use in a well-ventilated area.

SAFETY GLASSES

SOLVENT GLOVES

Sodium Chloride

[7647–14–5]

NaCl
Salt

UNCLASSIFIED

HAZARDS

Solid   Can be mildly irritating to skin and eyes as a solid or concentrated solution.

Dust   Can be mildly irritating to eyes and lungs.

0 / 1 / 0 (Manufacturer)

PRECAUTIONS

Avoid skin and eye contact. Use acid or solvent gloves and safety glasses. Use in a well-ventilated area.

ACID GLOVES   SAFETY GLASSES

Sodium Fluoride

[7681–49–4]

NaF

POISON

HAZARDS

Dust   May cause nausea, vomiting, headache, dizziness, and diarrhea. High concentrations may cause death.

IDLH   250 mg/m^3 (as fluorine)

POISON

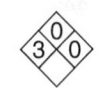

PRECAUTIONS

Avoid skin and eye contact. Use acid or solvent gloves and safety glasses. Use only within the confines of an air-exhausted hood.

ACID GLOVES   SAFETY GLASSES   AIR EXHAUSTED

Sodium Hydroxide

[1310–73–2]

NaOH
Caustic Soda
Lye

CORROSIVE (BASE)

HAZARDS

Solid/Liquid Can cause severe burns to skin and eyes.

Dust/Mist Can be highly irritating to skin, eyes, nose, throat, and lungs.

TLV/PEL 2 mg/m^3 C

IDLH 10 mg/m^3

CORROSIVE

0 / 3 / 1

PRECAUTIONS

Avoid skin and eye contact. Use acid gloves, faceshield and safety glasses, apron and armguards.
Use in a well-ventilated area.

FACESHIELD AND GLASSES   ACID GLOVES   APRON AND ARMGUARDS

369

Sodium Hypochlorite

[7681–52–9]

NaClO
Bleach

UNCLASSIFIED

HAZARDS

Solid/Liquid Can be slightly irritating to skin and eyes.

Dust/Mist Can be irritating to eyes, nose and throat.

Important Hazard The anhydrous salt is highly explosive; can react with acids to form chlorine gas in poisonous concentrations; can react violently with amines, ammonium acetate, ammonium oxalate, cellulose, and ethylene imine.

SPECIAL HAZARD

(5% solution, Manufacturer)

PRECAUTIONS

Avoid skin and eye contact. Use acid or solvent gloves and safety glasses. Use in a well-ventilated area.

ACID GLOVES   SAFETY GLASSES

Sodium Hypophosphite [7681–53–0]

NaH_2PO_5 **UNCLASSIFIED**

HAZARDS

Solid   Can be slightly irritating to skin and eyes.

Dust   Can be mildly irritating to eyes and lungs.

Important Hazard   Can explode on rapid heating. Decomposes to phosphine on heating.

SPECIAL HAZARD

(Manufacturer)

PRECAUTIONS

Avoid skin and eye contact. Use acid or solvent gloves and safety glasses.
Use in a well-ventilated area.
Keep away from heat.

ACID GLOVES   SAFETY GLASSES

Sodium Metasilicate

[6834–92–0]

CORROSIVE (BASE)

Na_2SiO_3
Sodium Silicate

HAZARDS

Solid/Liquid Can cause severe burns to skin and eyes.

Dust/Mist Can cause severe burns to eyes. May cause severe irritation of nose, throat, and lungs.

CORROSIVE

(Manufacturer)

PRECAUTIONS

Avoid skin and eye contact. Use acid gloves and safety glasses.
Use in a well-ventilated area.

ACID GLOVES SAFETY GLASSES

Sodium Nitrate

[7631–99–4]

NaNO$_3$ **OXIDIZER**

HAZARDS

Dust May irritate nose, throat, and lungs. May cause sore throat, coughing, and shortness of breath.

Important Hazard Explodes when heated to over 1000°F.

OXIDIZER

SPECIAL HAZARD

(Manufacturer)

PRECAUTIONS

Avoid skin and eye contact. Use acid or solvent gloves and safety glasses. Use in a well-ventilated area.

ACID GLOVES

SAFETY GLASSES

Sodium Nitrite

[7632–00–0]

NaNO$_2$

OXIDIZER

HAZARDS

Dust Can irritate nose, throat, and lungs. May cause sore throat, coughing, and shortness of breath. High concentrations may cause nausea, vomiting, and blood effects.

Important Hazard Explodes when heated to over 1000°F.

OXIDIZER

SPECIAL HAZARD

PRECAUTIONS

Avoid skin and eye contact. Use acid or solvent gloves and safety glasses. Use in a well-ventilated area.

ACID GLOVES

SAFETY GLASSES

Sodium Sulfate

[7757–82–6]

UNCLASSIFIED

Na$_2$SO$_4$
Disodium Sulfate

HAZARDS

Dust Repeated contact may cause skin irritation. May cause eye, nose, throat, and lung irritation.

PRECAUTIONS

Avoid skin and eye contact. Use solvent or acid gloves and safety glasses. Use in a well-ventilated area.

SAFETY GLASSES

SOLVENT GLOVES

Sodium Sulfite

[7757–83–7]

Na_2SO_3

CORROSIVE (BASE)

HAZARDS

Solid/Liquid Can be mildly irritating to skin and eyes.

Dust/Mist Can be irritating to eyes, nose, and throat.

Important Hazard Emits very toxic fumes of sodium monoxide and oxides of sulfur when heated to decomposition or when it comes into contact with acids.

CORROSIVE   SPECIAL HAZARD

(Manufacturer)

PRECAUTIONS

Avoid skin and eye contact. Use acid gloves and safety glasses.
Use in a well-ventilated area.

ACID GLOVES   SAFETY GLASSES

Sodium Tetraborate

[1303-43-4]

UNCLASSIFIED

$Na_2B_4O_7$
Sodium Borate
Boratets, Tetra, Sodium Salts

HAZARDS

Dust/Mist Can cause skin and eye irritation. Can cause intoxication, vomiting, and diarrhea at high concentrations.

TLV/PEL 1 mg/m^3 (anhydrous)
 5 mg/m^3 (decahydrate)
 1 mg/m^3 (pentahydrate)

(Pentahydrate, Manufacturer)

PRECAUTIONS

Avoid skin and eye contact. Use acid or solvent gloves and safety glasses. Use in a well-ventilated area.

ACID GLOVES   SAFETY GLASSES

Sodium Thiosulfate

[7772–98–7]

$Na_2S_2O_3$

UNCLASSIFIED

HAZARDS

Solid/Liquid Repeated contact may cause irritation of skin and eyes.

Dust/Mist High concentrations may be irritating to eyes, nose and throat.

Important Hazard Emits very toxic fumes of sodium monoxide and oxides of sulfur when heated to decomposition.

SPECIAL HAZARD

(Manufacturer)

PRECAUTIONS

Avoid skin and eye contact. Use acid or solvent gloves and safety glasses. Use in a well-ventilated area.

ACID GLOVES   SAFETY GLASSES

Solder Flux, Rosin   UNCLASSIFIED

HAZARDS

Liquid Rosin solder fluxes consist of about 90% resin acids and 10% neutral material. When fluxes contact hot solder, gases and particles form. These pyrolysis products include acetone, methyl alcohol, aliphatic aldehydes, carbon dioxide, carbon monoxide, methane, ethane, abietic acid, and other related diterpene acids.

Pyrolysis Products Strong odor when aliphatic aldehydes, such as formaldehyde, are present at a concentration of 0.07 mg/m^3. Can be irritating to eyes, nose, and throat.

TLV/PEL Sensitizer; reduce exposure to as low as possible (thermal decomposition products, as resin acids—colophony)

Important Hazard May cause an allergic skin rash in certain sensitive individuals.

SPECIAL HAZARD

PRECAUTIONS

Avoid skin and eye contact. Use solvent gloves, faceshield and safety glasses, apron and armguards when handling liquid. Use only within the confines of an air-exhausted hood for large soldering operations such as wave soldering.

FACESHIELD AND GLASSES   SOLVENT GLOVES   APRON AND ARMGUARDS   AIR EXHAUSTED

Stannic Chloride

[7646–78–8]

SnCl$_4$
Tin Chloride

CORROSIVE (ACID)

HAZARDS

Solid/Liquid Can cause chemical burns to skin and eyes.

Dust/Mist Can cause irritation of the nose, throat, and lungs. May cause nausea, vomiting, and diarrhea. Repeated exposure may cause breathing difficulties.

TLV/PEL 2 mg/m^3 (as tin)

VP 20 mm @ 22°C

Important Hazard Reacts with heat or water to form hydrochloric acid.

CORROSIVE   SPECIAL HAZARD

3 0 1

PRECAUTIONS

Avoid skin and eye contact. Use acid or solvent gloves, faceshield and safety glasses, apron and armguards.
Use only within the confines of an air-exhausted hood.

FACESHIELD AND GLASSES   ACID GLOVES   APRON AND ARMGUARDS   AIR EXHAUSTED

Stannous Chloride

[7772–99–8]

SnCl$_2$
Tin Dichloride

CORROSIVE (ACID)

HAZARDS

Solid/Liquid   Can cause skin and eye irritation.

Dust/Mist   Can cause irritation of the nose, throat, and lungs. May cause nausea, vomiting, and diarrhea. Repeated exposure can cause breathing difficulties.

TLV/PEL   2 mg/m^3 (as tin)

CORROSIVE

(Manufacturer)

PRECAUTIONS

Avoid skin and eye contact. Use acid gloves, faceshield and safety glasses, apron and armguards.
Use in a well-ventilated area.

FACESHIELD AND GLASSES   ACID GLOVES   APRON AND ARMGUARDS

Stannous Sulfate

[7488–55–3]

$SnSO_4$
Tin Sulfate

OXIDIZER

HAZARDS

Solid/Liquid Can cause chemical burns to skin and eyes.

Dust/Mist Can cause irritation of the nose, throat, and lungs. May cause nausea, vomiting, diarrhea, and breathing difficulties.

TLV/PEL 2 mg/m^3 (as tin)

OXIDIZER

(Manufacturer)

PRECAUTIONS

Avoid skin and eye contact. Use acid or solvent gloves, faceshield and safety glasses, apron and armguards.
Use in a well-ventilated area.

FACESHIELD AND GLASSES   ACID GLOVES   APRON AND ARMGUARDS

Stoddard Solvent

[8052–41–3]

C_7-C_{12} Hydrocarbons
Petroleum Distillate
Mineral Spirits

COMBUSTIBLE

HAZARDS

Liquid Can cause eye irritation. Prolonged or repeated skin exposure can cause drying or cracking due to defatting action.

Vapor Odor threshold is 0.9 ppm. Can be irritating to eyes, nose, and throat and can cause headache, tiredness, and giddiness at exposures of 100 ppm for 7 hours with previous exposure. Without previous exposure, concentrations of 200 to 400 ppm are needed to produce the same effects. Can also cause memory effects and liver and kidney damage. Combustible.

TLV/PEL	100 ppm	**FP**	100 – 110°F
IDLH	20,000 mg/m^3	**AT**	450°F
VP	2 mm (approx.)	**ER**	1.1 – 6%

COMBUSTIBLE

PRECAUTIONS

Avoid skin and eye contact. Use solvent gloves, faceshield and safety glasses, apron and armguards. Use only within the confines of an air-exhausted hood.

Keep away from open flames and high temperatures.

FACESHIELD AND GLASSES   SOLVENT GLOVES   APRON AND ARMGUARDS   AIR EXHAUSTED   NO IGNITION SOURCE

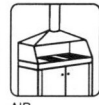

Styrene, Monomer

[100–42–5]

$C_6H_5CHCH_2$

FLAMMABLE

HAZARDS

Liquid Can be readily absorbed through the skin. Repeated contact can cause skin and eye irritation.

Vapor Odor detectable at 0.02 to 1.9 ppm. Can be irritating to eyes and nose. 200 to 700 ppm can cause drowsiness, headache, nausea, dizziness, and fatigue. May cause liver and kidney damage. Flammable.

TLV/PEL	20 ppm	**FP**	87°F
STEL	40 ppm	**AT**	914°F
IDLH	700 ppm	**ER**	1.1 – 7.0%
VP	5 mm		

Important Hazard Suspected carcinogen.

FLAMMABLE   CANCER CAUSING

PRECAUTIONS

Particular care should be taken to avoid skin and eye contact. Use solvent gloves, faceshield and safety glasses, apron and armguards. Use only within the confines of an air–exhausted hood.

Keep away from ignition sources such as flame, spark, and heat.

FACESHIELD AND GLASSES   SOLVENT GLOVES   APRON AND ARMGUARDS   AIR EXHAUSTED   NO IGNITION SOURCE

Succinic Acid

[110–15–6]

HOCOCH$_2$CH$_2$COOH

CORROSIVE (ACID)

HAZARDS

Solid Can cause chemical burns to skin and eyes.

Dust May cause irritation of eyes, nose, throat, lungs, and skin.

Important Hazard Can be an explosion hazard when heated.

CORROSIVE — SPECIAL HAZARD

(Manufacturer)

PRECAUTIONS

Avoid skin and eye contact. Use solvent gloves, faceshield and safety glasses, apron and armguards. Use in a well-ventilated area.
Keep away from flame and high temperatures.

FACESHIELD AND GLASSES — SOLVENT GLOVES — APRON AND ARMGUARDS — NO IGNITION SOURCE

Sulfamic Acid

[5329-14-6]

H_3NO_3S

CORROSIVE (ACID)

HAZARDS

Liquid Can cause chemical burns to skin and eyes.

Dust Can cause irritation of eyes, nose, throat, lungs, and skin.

Important Hazard Emits very toxic fumes of oxides of sulfur and oxides of nitrogen when heated to decomposition.

CORROSIVE   SPECIAL HAZARD   (Manufacturer)

PRECAUTIONS

Avoid skin and eye contact. Use solvent gloves, faceshield and safety glasses, apron and armguards.
Use only within the confines of an air-exhausted hood.

FACESHIELD AND GLASSES   SOLVENT GLOVES   APRON AND ARMGUARDS   AIR EXHAUSTED

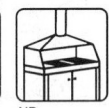

Sulfur Hexafluoride

[2551–62–4]

SF_6

NONFLAMMABLE GAS

HAZARDS

Gas If no significant amounts of other sulfur fluoride compounds are present as impurities, acts as a simple asphyxiant by displacing oxygen in the air. Extremely high concentrations can cause headache and dizziness.

TLV/PEL 1000 ppm

(Manufacturer) 0-1-0

PRECAUTIONS

Use in a well-ventilated area.

Note: This compound contributes to global warming if released to the atmosphere.

SAFETY GLASSES

Sulfuric Acid

[7664–93–9]

H_2SO_4

CORROSIVE (ACID)

HAZARDS

Liquid Can cause burns to skin and eyes. Repeated contact with dilute solutions can cause skin rash.

Mist Low concentrations can be mildly irritating to eyes and lungs. High concentrations can cause severe chemical burns to skin, eyes, and lungs. Also, erosion and discoloration of the teeth can occur. Repeated or prolonged inhalation can cause an inflammation of the lungs leading to chronic bronchitis.

TLV/PEL 1 mg/m^3 **IDLH** 15 mg/m^3

STEL 3 mg/m^3 **VP** < 0.001 mm

Important Hazard Sulfuric acid mists are a suspected carcinogen.

CORROSIVE   CANCER CAUSING

PRECAUTIONS

Avoid skin and eye contact. Use acid gloves, faceshield and safety glasses, apron and armguards. Use only within the confines of an air-exhausted hood. Mix slowly in small amounts when adding sulfuric acid to other liquids. When mixing sulfuric acid with hydrogen peroxide, add the hydrogen peroxide to the sulfuric acid.

FACESHIELD AND GLASSES   ACID GLOVES   APRON AND ARMGUARDS   AIR EXHAUSTED

Tellurium

[13494–80–9]

Te **UNCLASSIFIED**

HAZARDS

Dust/Fume Can cause a garlic odor of breath and sweat, dryness of the mouth, metallic taste, sleepiness, loss of appetite and nausea. Repeated exposure can cause lung irritation, damage to the blood cells and central nervous system depression.

TLV/PEL 0.1 mg/m^3

IDLH 25 mg/m^3

PRECAUTIONS

Avoid skin and eye contact. Use acid or solvent gloves and safety glasses. Use only within the confines of an air-exhausted hood.

| ACID GLOVES | SAFETY GLASSES | AIR EXHAUSTED |

389

Tergitol

$C_{36}H_{66}O_{10}$ **UNCLASSIFIED**

HAZARDS

Liquid May cause skin and eye irritation.

Mist Little indication of toxicity. High concentrations may cause nausea, vomiting, headache, dizziness, and diarrhea.

FP 311 – 729°F

PRECAUTIONS

Avoid skin and eye contact. Use solvent gloves and safety glasses.
Use in a well-ventilated area.

SAFETY GLASSES

SOLVENT GLOVES

Tetrachlorobutadiene

[921–09–5]

$C_4H_2Cl_4$

UNCLASSIFIED

HAZARDS

Liquid Can be readily absorbed through the skin. Repeated contact can cause skin and eye irritation.

Vapor Can cause eye and nose irritation and kidney damage.

Important Hazard Suspected carcinogen.

CANCER CAUSING

PRECAUTIONS

Tetrachlorobutadiene is present in the residues of some plasma aluminum etchers. Etcher residues should be kept within the confines of an air-exhausted hood or used with local exhaust ventilation.

Use solvent gloves, faceshield and safety glasses, apron and armguards. Particular care should be taken to avoid skin contact with the liquid and inhalation of vapors.

FACESHIELD AND GLASSES SOLVENT GLOVES APRON AND ARMGUARDS AIR EXHAUSTED

Tetrachloroethylene

[127–18–4]

UNCLASSIFIED

C_2Cl_4
Perchloroethylene
PERC

HAZARDS

Liquid Can cause eye irritation. Repeated contact can cause skin irritation.

Vapor Odor detectable at 5 to 50 ppm. Exposure to 100 to 200 ppm can cause irritation to the eyes, nose, and throat. At 200 to 300 ppm headaches, dizziness, and impaired judgment can also occur. High concentrations can cause liver and kidney damage.

TLV/PEL 25 ppm

STEL 100 ppm

IDLH 150 ppm

VP 16 mm

Important Hazard Suspected carcinogen.

CANCER CAUSING

PRECAUTIONS

Avoid skin and eye contact. Use solvent gloves, faceshield and safety glasses, apron and armguards. Use only within the confines of an air-exhausted hood.

FACESHIELD AND GLASSES   SOLVENT GLOVES   APRON AND ARMGUARDS   AIR EXHAUSTED

Tetraethyl Ortho Silicate [78–10–4]

$Si(OC_2H_5)_4$
Ethyl Silicate
Tetraethyl Silicate
TEOS

COMBUSTIBLE

HAZARDS

Liquid Can cause eye irritation. Prolonged or repeated skin exposure can cause drying or cracking due to defatting action.

Vapor Mild, sweet, alcohol-like odor detectable at 3.6 to 5 ppm. 250 ppm can be irritating to eyes and nose. 1200 ppm can cause tears to form. Repeated exposure to high concentrations may cause kidney, liver, and lung damage. Combustible.

TLV/PEL	10 ppm	**FP**	125°F
IDLH	700 ppm	**AT**	500°F
VP	2 mm	**ER**	0.9 – 7.2%

COMBUSTIBLE

PRECAUTIONS

Avoid skin and eye contact. Use solvent gloves, faceshield and safety glasses, apron and armguards. Use only within the confines of an air-exhausted hood.

Keep away from open flame and high temperatures.

FACESHIELD AND GLASSES   SOLVENT GLOVES   APRON AND ARMGUARDS   AIR EXHAUSTED   NO IGNITION SOURCE

393

Tetrahydrofuran

[109–99–9]

FLAMMABLE

$(C_2H_4)_2O$
THF
Diethylene Oxide

HAZARDS

Liquid May cause skin and eye irritation.

Vapor Odor detectable at 0.09 to 61 ppm. Can cause irritation to the eyes, nose, and throat. High concentrations can cause headaches, dizziness, and impaired judgment. Flammable.

TLV/PEL	200 ppm	**FP**	1.4°F
STEL	250 ppm	**AT**	610°F
IDLH	2000 ppm	**ER**	1.8 – 11.8%
VP	145 mm		

FLAMMABLE

PRECAUTIONS

Avoid skin and eye contact. Use solvent gloves, faceshield and safety glasses, apron and armguards. Use only within the confines of an air-exhausted hood.

Keep away from ignition sources such as flame, spark, and heat.

FACESHIELD AND GLASSES   SOLVENT GLOVES   APRON AND ARMGUARDS   AIR EXHAUSTED   NO IGNITION SOURCE

Tetramethyl Ammonium Hydroxide

[75–59–2]

$(CH_3)_4NOH$

POISON/ CORROSIVE (BASE)

HAZARDS

Solid/Liquid Can cause burns to skin and eyes. May be fatal if swallowed or absorbed through the skin.

Dust/Mist Can be irritating to eyes, skin, nose, and throat. High concentrations may cause death from fluid in lungs.

POISON   CORROSIVE

PRECAUTIONS

Avoid skin and eye contact. Use solvent gloves, faceshield and safety glasses, apron and armguards. Use only within the confines of an air-exhausted hood.

FACESHIELD AND GLASSES   SOLVENT GLOVES   APRON AND ARMGUARDS   AIR EXHAUSTED

Thionyl Chloride

[7719-09-7]

Cl_2SO
TCA
Sulfurous Oxychloride

CORROSIVE (ACID)

HAZARDS

Liquid Can cause severe skin and eye irritation. May cause chemical burns. May be fatal if swallowed or absorbed through the skin.

Vapor Pungent odor similar to sulfur dioxide. May cause eye irritation, coughing, burning of the throat, and a choking sensation.

TLV/PEL 1 ppm C

VP 100 mm at 21°C

Important Hazard Will react with water to form hydrogen chloride and sulfur dioxide.

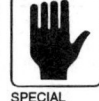

CORROSIVE   SPECIAL HAZARD

PRECAUTIONS

Particular care should be taken to avoid skin and eye contact. Use acid gloves, faceshield and safety glasses, apron and armguards. Use only within the confines of a closed, well-exhausted system.
Keep away from water, including small amounts of moisture.

FACESHIELD AND GLASSES   ACID GLOVES   APRON AND ARMGUARDS   AIR EXHAUSTED   WATER REACTIVE

Thiourea

[62–56–6]

NH_2CSNH_2

POISON

HAZARDS

Solid/Liquid May cause sores on skin and discoloration. May cause an allergic skin rash. May be fatal if swallowed.

Dust/Mist May damage blood and liver. May cause an enlargement of the thyroid gland.

Important Hazard Suspected carcinogen.

POISON

CANCER CAUSING

(Manufacturer)

PRECAUTIONS

Avoid skin and eye contact. Use acid or solvent gloves, faceshield and safety glasses, apron and armguards.
Use only within the confines of an air-exhausted hood.

FACESHIELD AND GLASSES

ACID GLOVES

APRON AND ARMGUARDS

AIR EXHAUSTED

Tin, Inorganic

[7440–31–5]

Sn

UNCLASSIFIED

HAZARDS

Dust/Fume Can cause irritation of eyes and skin. Can cause breathing difficulties and can settle in the lungs (stannosis).

TLV/PEL 2 mg/m^3

IDLH 100 mg/m^3

PRECAUTIONS

Avoid skin and eye contact. Use acid or solvent gloves and safety glasses. Use in a well-ventilated area.

ACID GLOVES   SAFETY GLASSES

Titanium

[7440–32–6]

Ti **UNCLASSIFIED**

HAZARDS

Dust In very high concentrations may cause irritation of the eyes, nose, throat, and skin. Exposure to titanium welding fumes may cause nausea, tightness of the chest, fever, and metallic taste.

Important Hazard The finely divided metal powder is flammable.

SPECIAL HAZARD

PRECAUTIONS

Avoid skin and eye contact. Use acid or solvent gloves and safety glasses.
Use in a well-ventilated area.
Do not use water or carbon dioxide to extinguish fires.

SAFETY GLASSES

SOLVENT GLOVES

Toluene

[108–88–3]

$C_6H_5CH_3$
Methylbenzene

FLAMMABLE

HAZARDS

Liquid Repeated contact can cause skin and eye irritation.

Vapor Sweet, "airplane glue"-like odor detectable at 0.2 to 2 ppm. Exposures of 50 to 100 ppm may cause fatigue, dizziness, and eye irritation. 200 to 500 ppm may cause symptoms similar to drunkenness, numbness, and nausea. Over 500 ppm may cause mental confusion and incoordination. High concentrations can cause liver damage. Flammable.

TLV/PEL	50 ppm	**FP**	40°F
STEL	150 ppm	**AT**	896°F
IDLH	500 ppm	**ER**	1.2 – 7%
VP	22 mm		

FLAMMABLE

PRECAUTIONS

Avoid skin and eye contact. Use solvent gloves, faceshield and safety glasses, apron and armguards. Use only within the confines of an air-exhausted hood.
Keep away from ignition sources such as flame, spark, and heat.

FACESHIELD AND GLASSES   SOLVENT GLOVES   APRON AND ARMGUARDS   AIR EXHAUSTED   NO IGNITION SOURCE

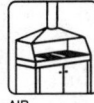

Toluene-2,4-Diisocyanate [584–84–9]

$CH_3C_6H_3(NCO)_2$
TDI

POISON/COMBUSTIBLE

HAZARDS

Liquid Can cause skin and eye irritation.

Vapor Medicated bandage-like odor detectable at 0.05 to 0.4 ppm. Can be irritating to eyes, nose, throat and lungs. Can cause choking sensation, difficult breathing, nausea, and fluid in lungs. Combustible.

TLV/PEL	0.005 ppm	**VP**	0.025 mm
STEL	0.02 ppm	**FP**	260°F
IDLH	2.5 ppm	**ER**	0.9 – 9.5%

Important Hazard Small exposures can cause allergic sensitization of the lungs.

POISON   COMBUSTIBLE   SPECIAL HAZARD

PRECAUTIONS

Avoid skin and eye contact. Use solvent gloves, faceshield and safety glasses, apron and armguards. Use only within the confines of an air-exhausted hood.
Keep away from open flame and high temperatures.
Keep away from water.

FACESHIELD AND GLASSES   SOLVENT GLOVES   APRON AND ARMGUARDS   AIR EXHAUSTED   NO IGNITION SOURCE   WATER REACTIVE

401

Trichloroacetic Acid

[76–03–9]

Cl$_3$CCOOH
TCA

**POISON/
CORROSIVE (ACID)**

HAZARDS

Liquid Can cause severe skin and eye irritation. May cause chemical burns. May be fatal if swallowed.

Vapor/Mist Sharp odor detectable at 0.2 to 0.4 ppm. May cause headache, nausea, dizziness, fatigue, and weakness in the arms and legs. May cause coughing, chest pains, nausea, and vomiting. Extreme exposures may cause shortness of breath and a life-threatening build-up of fluid in lungs that may not appear until several hours after exposure.

TLV/PEL 1 ppm

VP 1 mm at 51°C

POISON   CORROSIVE

(Manufacturer)

PRECAUTIONS

Avoid skin and eye contact. Use solvent gloves, faceshield and safety glasses, apron and armguards. Use only within the confines of an air-exhausted hood.

FACESHIELD AND GLASSES   SOLVENT GLOVES   APRON AND ARMGUARDS   AIR EXHAUSTED

Trichlorobenzene

[120–82–1]

$C_6H_3Cl_3$

COMBUSTIBLE

HAZARDS

Liquid Repeated contact can cause skin and eye irritation.

Vapor Odor detectable at 3 ppm. Can cause irritation of eyes, nose, and throat. Prolonged exposure at high concentrations may cause liver damage. Combustible.

TLV/PEL	5 ppm C	**AT**	1060°F
VP	1 mm at 38.4°C	**ER**	2.5 – 6.6% @ 150°C
FP	225 – 235°F		

COMBUSTIBLE

PRECAUTIONS

Avoid skin and eye contact. Use solvent gloves, faceshield and safety glasses, apron and armguards. Use only within the confines of an air-exhausted hood.

Keep away from open flame and high temperatures.

FACESHIELD AND GLASSES   SOLVENT GLOVES   APRON AND ARMGUARDS   AIR EXHAUSTED   NO IGNITION SOURCE

1,1,1-Trichloroethane

[71–55–6]

CH₃CCl₃
TCA
Methyl Chloroform

UNCLASSIFIED

HAZARDS

Liquid Repeated contact can cause skin and eye irritation.

Vapor Odor detectable at 16 to 100 ppm. Can be irritating to eyes and lungs. Incoordination and impaired judgment can occur at exposures of 500 to 1000 ppm. Does not burn at ordinary temperatures, but elevated temperatures — such as from a cutting torch — can ignite the material or form toxic gases.

TLV/PEL	350 ppm	**VP**	100 mm
STEL	450 ppm	**AT**	932°F
IDLH	700 ppm	**ER**	7 – 16%

PRECAUTIONS

Avoid skin and eye contact. Use solvent gloves, faceshield and safety glasses, apron and armguards. Use only within the confines of an air-exhausted hood.

FACESHIELD AND GLASSES SOLVENT GLOVES APRON AND ARMGUARDS AIR EXHAUSTED

Trichloroethylene

[79–01–6]

C_2Cl_3H
TCE

UNCLASSIFIED

HAZARDS

Liquid Can cause eye irritation. Repeated contact can cause skin irritation.

Vapor Odor detectable at 0.5 to 167 ppm. Can cause headache, dizziness, sleepiness, and irritability. Exposure at 100 to 600 ppm may cause dizziness, headache, vertigo, nausea, and fatigue. Exposure to 1000 ppm for 2 hours can cause visual and coordination problems. Repeated exposure can cause liver and kidney damage. Vapors can be moderately flammable at high temperatures.

TLV/PEL	50 ppm	**VP**	58 mm
STEL	100 ppm	**AT**	770°F
IDLH	1000 ppm	**ER**	8.0 – 10.5% @ 25°C

Important Hazard Suspected carcinogen.

CANCER CAUSING

PRECAUTIONS

Avoid skin and eye contact. Use solvent gloves, faceshield and safety glasses, apron and armguards. Use only within the confines of an air-exhausted hood.

FACESHIELD AND GLASSES   SOLVENT GLOVES   APRON AND ARMGUARDS   AIR EXHAUSTED

Trichlorosilane

[10025–78–2]

SiHCl₃

FLAMMABLE/ CORROSIVE (ACID)

HAZARDS

Liquid Can cause chemical burns to skin and eyes.

Vapor Can be very irritating and cause chemical burns to eyes and lungs. Fumes in presence of moist air.

VP 400 mm at 14.5°C

FP –18°F

AT 219°F

Important Hazard Extremely flammable. Can be spontaneously flammable in air. Reacts with water to form hydrogen chloride (see Hydrogen Chloride).

FLAMMABLE   SPECIAL HAZARD   AIR REACTIVE

PRECAUTIONS

Use safety glasses. Use only within a closed, well-exhausted system. Keep away from ignition sources such as flame, spark, and heat.
Keep away from water.

SAFETY GLASSES   AIR EXHAUSTED   NO IGNITION SOURCE   WATER REACTIVE

Triethanolamine

[102–71–6]

$(HOC_2H_4)_3N$
TEA

COMBUSTIBLE

HAZARDS

Liquid May cause skin and eye irritation. Repeated contact may cause an allergic skin rash in some individuals.

Mist May be irritating to eyes, nose, and throat. Can cause coughing. Combustible.

TLV/PEL 0.5 ppm

VP < 0.01 mm

FP 354°F

AT 600°F (approx.)

ER 1.3 – 8.5% (approx.)

Important Hazard Suspected carcinogen.

COMBUSTIBLE   CANCER CAUSING

PRECAUTIONS

Avoid skin and eye contact. Use solvent gloves, faceshield and safety glasses, apron and armguards. Use with adequate ventilation.
Keep away from open flame and high temperatures.

FACESHIELD AND GLASSES   SOLVENT GLOVES   APRON AND ARMGUARDS   AIR EXHAUSTED   NO IGNITION SOURCE

Triethylene Glycol

[112–27–6]

$C_6H_{14}O_8$
Triglycol
TEG

COMBUSTIBLE

HAZARDS

Liquid May cause skin and eye irritation.

Mist May be irritating to eyes and lungs. Combustible.

VP < 0.001 mm

FP 350°F

AT 700°F

ER 0.9 – 9.2%

COMBUSTIBLE

1 / 1 / 0

PRECAUTIONS

Avoid skin and eye contact. Use solvent gloves, faceshield and safety glasses, apron and armguards. Use with adequate ventilation.
Keep away from open flame and high temperatures.

FACESHIELD AND GLASSES | SOLVENT GLOVES | APRON AND ARMGUARDS | NO IGNITION SOURCE

Trimethylaluminum [75–24–1]

(CH$_3$)$_3$Al **FLAMMABLE**

HAZARDS

Liquid Instant severe burns by contact.

Dust Reacted trimethylaluminum forms very fine particles which in large concentrations can cause metal fume fever.

VP 8.4 mm

Important Hazard Pyrophoric — ignites and burns on contact with air or water.

| FLAMMABLE |  AIR REACTIVE | SPECIAL HAZARD |

PRECAUTIONS

Avoid skin and eye contact. Use acid or solvent gloves, faceshield and safety glasses, apron and armguards. Use only within a closed, well-exhausted system.

Keep away from ignition sources such as flame, spark, and heat. Keep away from water.

Fire extinguishers are ineffective in controlling fires from gas leaks; the gas supply must be turned off.

| FACESHIELD AND GLASSES | ACID GLOVES | APRON AND ARMGUARDS | AIR EXHAUSTED | NO IGNITION SOURCE | WATER REACTIVE |

409

Trimethylantimony

[594–10–5]

(CH$_3$)$_3$Sb

FLAMMABLE/POISON

HAZARDS

Liquid May cause irritation of skin and eyes.

Vapor/Dust Reacted trimethylantimony forms very fine particles of antimony trioxide. May cause irritation of the eyes, nose, throat, and lungs. May cause stomach upset, irritability, dizziness, and muscular pain. Vapors are flammable.

TLV/PEL 0.5 mg/m^3 (as antimony)   **VP** 78.6 mm

Important Hazard May be pyrophoric — ignites and burns on contact with air or water.

| FLAMMABLE | AIR REACTIVE | POISON | SPECIAL HAZARD |

PRECAUTIONS

Avoid skin and eye contact. Use acid or solvent gloves, faceshield and safety glasses, apron and armguards. Use only within a closed, well-exhausted system.

Keep away from ignition sources such as flame, spark, and heat. Keep away from water.

Fire extinguishers are ineffective in controlling fires from gas leaks; the gas supply must be turned off.

| FACESHIELD AND GLASSES | ACID GLOVES | APRON AND ARMGUARDS | AIR EXHAUSTED | NO IGNITION SOURCE | WATER REACTIVE |

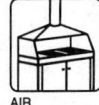

Trimethylarsenic

[593–88–4]

(CH$_3$)$_3$As

FLAMMABLE

HAZARDS

Liquid May cause chemical burns to skin and eyes.

Vapor Pungent odor. May cause irritation of the stomach and intestines, nausea, vomiting, diarrhea, cold sweats, weak rapid pulse, and loss of appetite.

Dust Reacted trimethylarsenic forms very fine particles of arsenic trioxide (see Arsenic Trioxide).

TLV/PEL 0.01 mg/m^3 (as arsenic)

VP 238 mm

Important Hazard Highly flammable; releases arsenic trioxide fumes on decomposition. Also, decomposes when contacted with air or moisture. Arsenic trioxide is a suspected carcinogen.

FLAMMABLE   AIR REACTIVE   CANCER CAUSING   SPECIAL HAZARD

PRECAUTIONS

Avoid skin and eye contact. Use acid or solvent gloves, faceshield and safety glasses, apron and armguards. Use only within a closed, well-exhausted system.

Keep away from ignition sources such as flame, spark, and heat. Keep away from water.

FACESHIELD AND GLASSES   ACID GLOVES   APRON AND ARMGUARDS   AIR EXHAUSTED   NO IGNITION SOURCE   WATER REACTIVE

Trimethyl Borate

[121–43–7]
FLAMMABLE

B(OCH$_3$)$_3$
Methyl Borate

HAZARDS

Liquid May cause skin and eye irritation. May be readily absorbed through the skin.

Vapor Can be irritating to the eyes, nose, and throat. Falmmable.

VP 100 mm

FP 30°F

PRECAUTIONS

Avoid skin and eye contact. Use solvent gloves, faceshield and safety glasses, apron and armguards. Use only within the confines of an air-exhausted hood.

Keep away from ignition sources such as flame, spark, and heat.

FACESHIELD AND GLASSES | SOLVENT GLOVES | APRON AND ARMGUARDS | AIR EXHAUSTED | NO IGNITION SOURCE

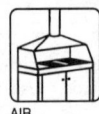

Trimethyl Phosphite

[121–45–9]

COMBUSTIBLE

P(OCH$_3$)$_3$
Methyl Phosphite

HAZARDS

Liquid Can cause skin and eye irritation.

Vapor Pungent, oily odor detectable at 0.001 ppm. Not generally considered objectionable until exposures approach 20 ppm. Can cause moderate to severe irritation of the eyes, nose, and throat. Combustible.

TLV/PEL 2 ppm

VP 24 mm @ 25°C

FP 130°F

COMBUSTIBLE

PRECAUTIONS

Avoid skin and eye contact. Use solvent gloves, faceshield and safety glasses, apron and armguards. Use in an air-exhausted hood.
Keep away from open flame and high temperatures.

FACESHIELD AND GLASSES | SOLVENT GLOVES | APRON AND ARMGUARDS | AIR EXHAUSTED | NO IGNITION SOURCE

Trisodium Phosphate [7601–54–9, Anhydrous]
[10101–89–0, Crystalline]

$Na_3PO_4 \cdot 12H_2O$

TSP

Trisodium Phosphate
　Dodecahydrate

UNCLASSIFIED

HAZARDS

Solid/Liquid Repeated exposure can cause skin irritation. Can cause eye irritation and damage.

Dust/Mist May cause irritation of nose and throat at exposures of 0.5 to 2.0 mg/m^3 for 1 hour. High concentrations may cause coughing and choking.

CORROSIVE

(Manufacturer)

PRECAUTIONS

Avoid skin and eye contact. Use acid gloves and safety glasses.
Use in a well-ventilated area.

ACID GLOVES　SAFETY GLASSES

Tungsten [7440–33–7]

W **UNCLASSIFIED**

HAZARDS

Dust Can cause skin and eye irritation. In very high concentrations may cause loss of appetite, incoordination of movement, trembling, and difficulty in breathing.

TLV/PEL 5 mg/m^3

STEL 10 mg/m^3

Important Hazard The finely divided metal powder is highly flammable.

SPECIAL HAZARD

(Manufacturer) 1/1/0

PRECAUTIONS

Avoid skin and eye contact. Use acid or solvent gloves and safety glasses. Use in a well-ventilated area.

SAFETY GLASSES

SOLVENT GLOVES

Tungsten Hexafluoride [7783-82-6]

WF_6

CORROSIVE (ACID)

HAZARDS

Liquid Can cause severe chemical burns to skin and eyes. Pain, redness, swelling, and early tissue death can occur.

Gas/Vapor Can cause severe burns to skin, eyes, and lungs. Low concentrations can cause headache, dizziness, labored breathing, excessive saliva secretion, cough, and chest pains. High concentrations can cause fluid in the lungs and damage to the lungs and eyes.

Important Hazard Reacts violently with water to produce oxyfluoride and hydrofluoric acids.

CORROSIVE   SPECIAL HAZARD

PRECAUTIONS

Avoid skin and eye contact. Use acid gloves, faceshield and safety glasses, apron and armguards. Use only within the confines of an air-exhausted hood or closed system. Keep away from moisture. Thoroughly clean and dry all equipment used in tungsten hexafluoride service.

FACESHIELD AND GLASSES   ACID GLOVES   APRON AND ARMGUARDS   AIR EXHAUSTED   WATER REACTIVE

Vinyl Cyclohexene Dioxide

[106–87–6]

$C_8H_{12}O_2$
Vinylhexane Dioxide

COMBUSTIBLE

HAZARDS

Liquid Can be readily absorbed through the skin. Can cause skin and eye irritation.

Vapor Can be irritating to eyes and lungs. High concentrations may cause congestion of the lungs. Combustible.

TLV/PEL 0.1 ppm

VP < 0.1 mm

FP 230°F

Important Hazard Suspected carcinogen.

COMBUSTIBLE CANCER CAUSING

PRECAUTIONS

Particular care should be taken to avoid skin and eye contact. Use solvent gloves, faceshield and safety glasses, apron and armguards. Use only within the confines of an air-exhausted hood.
Keep away from open flame and high temperatures.

FACESHIELD AND GLASSES   SOLVENT GLOVES   APRON AND ARMGUARDS   AIR EXHAUSTED   NO IGNITION SOURCE

Xylene

[1330–20–7]

C_8H_{10}
Xylol

FLAMMABLE

HAZARDS

Liquid Can be readily absorbed through the skin. Repeated contact can cause skin and eye irritation.

Vapor Sweet, "airplane glue"-like odor detectable at 0.08 to 40 ppm. Exposure at 200 ppm can cause eye, nose, and throat irritation. High concentrations can cause headache, dizziness, nausea, loss of appetite, and fatigue. Repeated exposure may cause liver and kidney damage. Flammable.

TLV/PEL	100 ppm	**FP**	81°F
STEL	150 ppm	**AT**	869 – 986°F
IDLH	900 ppm	**ER**	1 – 7%
VP	7 mm		

FLAMMABLE

PRECAUTIONS

Particular care should be taken to avoid skin and eye contact. Use solvent gloves, faceshield and safety glasses, apron and armguards. Use only within the confines of an air-exhausted hood.

Keep away from ignition sources such as flame, spark, and heat.

FACESHIELD AND GLASSES   SOLVENT GLOVES   APRON AND ARMGUARDS   AIR EXHAUSTED   NO IGNITION SOURCE

Zinc Oxide

[1314–13–2]

ZnO

UNCLASSIFIED

HAZARDS

Fume Can cause fever, chills, muscular pain, nausea, and vomiting.

Dust Nuisance dust. Can cause skin and eye irritation. Very high airborne concentrations may cause eye, nose, and throat irritation.

TLV/PEL	5 mg/m^3 (fume)
	10 mg/m^3 (total dust)
IDLH	500 mg/m^3
STEL	10 mg/m^3 (fume)

NFPA diamond: Health 1, Flammability 0, Reactivity 0 (Manufacturer)

PRECAUTIONS

Avoid skin and eye contact. Use acid or solvent gloves and safety glasses. Use in a well-ventilated area.

SAFETY GLASSES

SOLVENT GLOVES

12.0 References

Basic Radiation Protection Criteria (NCRP Report No. 39). National Council on Radiation Protection and Measurements, 7910 Woodmont Ave., Bethesda, MD 20814 (1971).

CA Selects: Chemical Hazards, Health and Safety. Chemical Abstract Service, 2540 Oleutangy River Rd., Columbus, OH 43210.

CCINFOdisc. Canadian Centre for Occupational Health and Safety, 250 Main St., East Hamilton, Ontario, Canada L8N 1H6.

Chemical Hazards of the Workplace (3d ed.), G.L. Hathaway, N.H. Proctor, J.P. Hughes, and M.L. Fischman. J. P. Lippincott Co., East Washington Square, Philadelphia, PA 19105 (1991).

29 Code of Federal Regulations: Parts 1900 to 1910. Occupational Safety and Health Administration, (1995).

Dangerous Properties of Industrial Materials (9th ed.), N. Irving Sax. Van Nostrand Reinhold Co., 115 Fifth Ave., New York, NY 10003 (1996).

Documentation of the Threshold Limit Values and Biological Exposure Indices (6th ed.). American Conference of Governmental Industrial Hygienists, Kemper Woods Center, 1330 Kemper Meadow Dr., Cincinnati, OH 45240 (1991).

Fire Protection Guide to Hazardous Materials (11th ed.). National Fire Protection Association, One Batterymarch Park, Quincy, MA 02269 (1994).

Genium MSDSs. Genium Publishing Corp., Room 407, One Genium Plaza, Schenectady, NY 12304-4690.

Handbook of Environmental Data on Organic Chemicals (2nd ed.). Van Nostrand Reinhold Co., 135 West 50th St., New York, NY 10020 (1983).

Hawley's Condensed Chemical Dictionary (12th ed.). Van Nostrand Reinhold Co., 135 West 50th St., New York, NY 10020 (1993).

IARC Monographs on the Evaluation of the Carcinogenic Risk of Chemicals. Geneva: World Health Organization, International Agency for Research on Cancer, 49 Sheridan Ave., Albany, NY 12210.

Knight-Ridder Information, Inc. (computer data bases), 2440 El Camino Real, Mountain View, CA 94040-1400.

Matheson Gas Data Book (6th ed.), William Braker and Allen L. Mossman. Matheson Division, Searle Medical Products USA, Inc., Lyndhurst, NJ 07071 (1980).

Merck Index: An Encyclopedia of Chemicals and Drugs. Merck and Co., Inc., 126 East Lincoln Ave., Rahway, NJ 07065 (1989).

NIOSH/OSHA Occupational Health Guidelines for Chemical Hazards. DHEW (NIOSH) Pub. No. 81-123 (1981).

NIOSH/OSHA Pocket Guide to Chemical Hazards. DHEW (NIOSH) Pub. No. 94-116 (1994).

Odor Thresholds for Chemicals with Established Occupational Health Standards. American Industrial Hygiene Association, 475 Wolf Ledges Parkway, Akron, OH 44311-1087 (1989).

Patty's Industrial Hygiene and Toxicology (4th ed.). G. D. Clayton and F. E. Clayton (Eds.). John Wiley & Sons, 605 Third Ave., New York, NY 10158-0012 (1994).

The Provisions of the Basic Safety Standards for Radiation Protection Relevant to the Protection of Workers Against Ionising Radiation. International Labour Office, CH-1211, Geneva 22, Switzerland (1985).

Radiation Frequency Electromagnetic Fields (NCRP Report No. 67). National Council on Radiation Protection and Measurements, 7910 Woodmont Ave., Bethesda, MD 20814 (1981).

Semiconductor Industrial Hygiene Handbook. M.E. Williams and D.G. Baldwin, Noyes Publications, 120 Mill Road, Park Ridge, NJ 07656 (1995).

Threshold Limit Values and Biological Exposure Indices for 1995–1996. American Conference of Governmental Industrial Hygienists, Kemper Woods Center, 1330 Kemper Meadow Dr., Cincinnati, OH 45240.

TOMES Plus. Micromedex, Inc., 6200 South Syracuse Way, Suite 300, Englewood, CO 80111-4740.

Uniform Fire Code — Article 51 and Article 80. Western Fire Chiefs Association, 5360 South Workman Mill Rd., Whittier, CA 90601 (1991).

Index for Chemical Hazards and Precautions

A

Acetic Acid **111**
Acetone **112**
Acetonitrile **113**
Acetylene **114**
Alkyl Benzenes **115**
Alumina 120
Aluminum, Metal **116**
Aluminum Acetate **117**
Aluminum Chloride **118**
Aluminum Fluoride **119**
Aluminum Oxide **120**
Aluminum Trifluoride 119
Ammate 130
Ammonia **121**
Ammonium Bichromate 125
Ammonium Bifluoride **122**
Ammonium Chloride **123**
Ammonium Citrate Dibasic **124**
Ammonium Dichromate **125**
Ammonium Fluoride **126**
Ammonium Hydroxide **127**
Ammonium Peroxydisulfate 128
Ammonium Persulfate **128**
Ammonium Phosphate **129**
Ammonium Sulfamate **130**
Anhydrous Ammonia 121
Anisole 294
Antimony **131**
Antimony Trioxide **132**
Aqua Regia **133**
Argon 258
Aromatic Hydrocarbons 115
Arsenic **134**
Arsenic Trioxide **135**
Arsine **136**
Asbestos **137**
Azabenzene 348
Azine 348

B

Baking Soda 363
Barium Chloride **138**
Barium Dinitrate 140
Barium Hydroxide **139**
Barium Nitrate **140**
Benzene Methanol **141**
Benzyl Alcohol 141
Beryllia 143
Beryllium **142**
Beryllium Oxide **143**
BGE 156
Bipotassium Chromate 332
Bleach 370
Boric Acid **144**
Boric Anhydride 146
Boroethane 192
Boron Hydride 192
Boron Nitride **145**
Boron Oxide **146**
Boron Tribromide **147**
Boron Trichloride **148**

Main entries are indicated by **boldface** page numbers.

Boron Trifluoride **149**
Boron Trioxide 146
Butane **150**
n-Butanol 155
2-Butanone 290
2-Butoxyethanol **151**
2-(2-Butoxyethoxy) Ethanol **152**
2-Butoxyethyl Acetate **153**
n-Butyl Acetate **154**
n-Butyl Alcohol **155**
Butyl Carbitol 152, 200
Butyl Cellosolve 151
Butyl Glycidyl Ether **156**
Butyrolactone **157**

C

Calcium Carbonate **158**
Calcium Chloride **159**
Calcium Hypochlorite **160**
Calcium Oxide **161**
Calcium Oxychloride 160
Calcium Sulfate **162**
Carbitol 201
Carbitol Acetate 202
Carbolic Acid 316
Carbon 163
Carbon Bisulfide 165
Carbon Black **163**
Carbon Dioxide **164**
Carbon Disulfide **165**
Carbon Monoxide **166**
Carbon Tetrachloride **167**
Carbon Tetrafluoride **168**
Carbowax 328
Catechol **169**
Caustic Soda 369
Chlorine **170**
Chlorobenzene **171**
Chloroform **172**
Chloromethane 287
o-Chlorotoluene **173**
Chromic Acid **174**

Chromic Sulfate **175**
Chromium, Hexavalent **176**
Chromium Sulfate 175
Chromium Trioxide 174
Chromous Salts 175
Citric Acid **177**
Coal Oil 264
Cobalt Nitrate **178**
Cobaltous Nitrate 178
Cobalt Sulfamate **179**
Copper **180**
Copper Chloride **181**
Copper Nitrate **182**
Copper Oxide **183**
Copper Sulfate **184**
Cresol **185**
Cupric Chloride 181
Cupric Nitrate 182
Cupric Oxide 183
Cupric Sulfate 184
Cyanide Salts **186**
Cyanogen **187**
Cyanogen Chloride **188**
Cyclohexane **189**
Cyclohexanone **190**

D

DEG 198
Diacetone 191
Diacetone Alcohol **191**
1,2-Diaminoethane 217
Diammonium Citrate 124
Diborane **192**
o-Dichlorobenzene **193**
p-Dichlorobenzene **194**
trans 1,2-Dichloroethylene **195**
Dichloromethane 289
Dichlorosilane **196**
Diesel Oil **197**
1-4-Diethylene Dioxide 207
Diethylene Glycol **198**
Diethylene Glycol Dimethyl Ether **199**

Diethylene Glycol Monobutyl Ether 152, **200**
Diethylene Glycol Monoethyl Ether **201**
Diethylene Glycol Monoethyl Ether Acetate **202**
Diethylene Oxide 297, 394
Diethylether 220
Diethyl Telluride **203**
Diglyme 199
Dimethyl Acetamide **204**
Dimethyl Formamide **205**
Dimethyl Ketone 112
Dimethyl Sulfoxide **206**
Dioxane **207**
Dipotassium Chromate 332
Dipropylene Glycol Methyl Ether **208**
Disodium Sulfate 375
DMSO 206
DPGME 208
Dross, Solder **209**

E

EDTA 218
EGMEA 213
Epoxy Butoxypropane 156
Epoxy Resins **210,** 351
1,2-Ethanediol 219
Ethanol 215
Ethanolamine **211**
Ether 220
2-Ethoxyethanol **212**
2-Ethoxyethyl Acetate **213**
Ethyl Acetate **214**
Ethyl Alcohol **215**
Ethyl Benzene **216**
Ethylene Alcohol 219
Ethylenediamine **217**
Ethylenediaminetetraacetic Acid **218**
Ethylene Dinitrilotetraacetic Acid 218
Ethylene Glycol **219**
Ethylene Glycol Monobutyl Ether Acetate 153

Ethylene Glycol Monoethyl Ether 212
Ethylene Glycol Monoethyl Ether Acetate 213
Ethylene Glycol Monomethyl Ether 282
Ethylene Glycol Monomethyl Ether Acetate 283
Ethylene Glycol Monopropyl Ether 262
Ethyl Ether **220**
2-Ethylhexanol 309
Ethyl Lactate **221**
Ethyl Silicate 393

F

Ferric Chloride **222**
Ferrous Chloride **223**
Ferrous Sulfate **224**
Fluoboric Acid, 48% **225**
Fluorine **226**
Fluoroboric Acid 225
Formaldehyde **227**
Formic Acid **228**
Freon-14 168
Freons **229**
Fuel Oil No. 2 197
Furfuryl Alcohol **230**

G

Gallium **231**
Gallium Arsenide **232**
Gasoline **233**
Germane **235**
Germanium **234**
Germanium Hydride 235
Germanium Tetrahydride **235**
Glacial Acetic Acid 111
Glutamic Acid Hydrochloride **236**
Glutaric Acid **237**
Glycerin **238**
Glycerol 238
Glycol Ether DE Acetate 202

H

Halons **239**
Helium 258
Hexachlorobutadiene **240**
Hexachloroethane **241**
Hexamethyldisilazane **242**
n-Hexane **243**
2-Hexanone 286
Hexone 292
Hexylene Glycol **244**
HMDS 242
Hydrazine **245**
Hydrazine Sulfate **246**
Hydrobromic Acid 248
Hydrochloric Acid 249
Hydrocyanic Acid 250
Hydrofluoric Acid 251
Hydrogen **247**
Hydrogen Bromide **248**
Hydrogen Chloride **249**
Hydrogen Cyanide **250**
Hydrogen Fluoride **251**
Hydrogen Peroxide **252**
Hydrogen Selenide **253**
Hydrogen Sulfide **254**
Hydrogen Tetrafluoroborate 225
Hydroquinone **255**

I

Indium **256**
Indium Phosphide **257**
Inert Gases **258**
Inorganic Lead and Tin 209
Iodine **259**
IPA 263
Iron Oxide **260**
Isobutene 261
Isobutylene **261**
Isopropanol 263
2-Isopropoxyethanol **262**
Isopropyl Alcohol 263

K

Kaowool 350
Kerosene **264**
Kerosine 264

L

Laughing Gas 307
Lead **265**
Lead Acetate **266**
Lead Nitrate **267**
Limestone 158
Limonene **268**
Liquefied Petroleum Gas **269**
Lithium Hydroxide **270**
LPG 269
Lye 336, 369

M

Magnesium **271**
Magnesium Acetate **272**
Magnesium Chloride **273**
Magnesium Dioxide **274**
Magnesium Oxide **275**
Manganous Sulfate **276**
MBK 286
MDI **288**
MEK 290
Melamine **277**
Mercuric Acetate **278**
Mercuric Oxide **279**
Mercury, Metal **280**
Mercury Acetate 278
Methane **281**
Methanoic Acid 228
Methanol 285
Methoxybenzene 294
2-Methoxyethanol **282**
2-Methoxyethyl Acetate **283**
1-Methoxy-2-Propanol 346
1-Methoxy-2-Propyl Acetate **284**, 347

Methylacetic Acid 344
Methyl Alcohol **285**
3-Methyl-2-Butanone 293
Methyl n-Butyl Ketone **286**
Methyl Chloride **287**
Methyl Chloroform 404
Methyl Cyanide 113
Methylene Bisphenyl Isocyanate **288**
Methylene Chloride **289**
Methylene Glycol 345
Methyl Ethyl Ketone **290**
5-Methyl-2-Hexanone 291
Methyl Isoamyl Ketone **291**
Methyl Isobutyl Ketone **292**
Methyl Isopropyl Ketone **293**
Methyl Pentanediol 244
Methyl Phenyl Ether **294**
Methyl Propyl Ketone **295**
N-Methyl-2-Pyrrolidone 298
Methyl Sulfoxide 206
MIAK 291
MIBK 292
Mineral Oil **296**
Mineral Spirits 383
MIPK 293
Monochlorobenzene 171
Monoethanolamine 211
Morpholine **297**
MPK 295
M-Pyrol 298

N

Naphtha **299**
Naphthalene **300**
NBA 154
Nickel **301**
Nitric Acid **302**
Nitric Oxide **303**
Nitrogen 258
Nitrogen Dioxide **304**
Nitrogen Monoxide 303
Nitrogen Oxide 307

Nitrogen Trifluoride **305**
Nitromethane **306**
Nitrous Oxide **307**
NMP 298
Novolac Resins 351

O

Octane **308**
Octanol, Mixed Isomers **309**
Octyl Alcohol 309
Oil Mist, Mineral 296
Oxalic Acid **310**
Oxygen **311**
Ozone **312**

P

Paraffin Oil 296
PEA 317
Pentane **313**
2-Pentanone 295
PERC 392
Perchloric Acid **314**
Perchloroethylene 392
Periodic Acid **315**
Petroleum Distillate 233, 383
Petroleum Distillates **299**
Petroleum Ether 299
PGME 346
PGMEA 284, 347
Phenethylamine 317
Phenol **316**
Phenolic Resins 351
Phenylethane 216
Phenylethylamine **317**
Phenyltrichlorosilane **318**
Phosphine **319**
Phosphoric Acid **320**
Phosphoric Anhydride 324
Phosphorus, Red **321**
Phosphorus Bromide 325
Phosphorus Oxychloride **322**

Phosphorus Pentafluoride 323
Phosphorus Pentoxide 324
Phosphorus Tribromide 325
α-Pinene 326
Platinum 327
PM Acetate 347
POCL 322
Polyethylene Glycol 328
Polyimide Resins 351
Polysiloxanes 356
Potassium Acid Phthalate 335
Potassium Bichromate 334
Potassium Biphthalate 335
Potassium Bromide 329
Potassium Carbonate 330
Potassium Chloride 331
Potassium Chromate 332
Potassium Cyanate 333
Potassium Dichromate 334
Potassium Hydrogen Phthalate 335
Potassium Hydroxide 336
Potassium Iodate 337
Potassium Iodide 338
Potassium Nitrate 339
Potassium Permanganate 340
Potassium Phosphate 341
Potassium Sulfate 342
Propane 343
1,2-Propanediol 345
Propionic Acid 344
1,2-Propylene Glycol 345
Propylene Glycol Monomethyl Ether 346
Propylene Glycol Monomethyl Ether Acetate 284, 347
Pyridine 348
Pyrocatechol 169
Pyrogallic Acid 349
Pyrogallol 349

Q

Quicklime 161
Quicksilver 280

R

Refractory Ceramic Fiber 350
Resins 351
RTV Silicone Rubber 356
Rubber Solvent 299

S

Salt 367
Selenium 352
Silane 353
Silica, Amorphous 354
Silicon 355
Silicon Chloride 357
Silicon Dioxide 354
Silicone 356
Silicon Tetrachloride 357
Silicon Tetrafluoride 358
Silicon Tetrahydride 353
Silver 359
Silver Acetate 360
Silver Nitrate 361
Soda Ash 366
Sodium Acetate 362
Sodium Bicarbonate 363
Sodium Bisulfate 364
Sodium Bisulfite 365
Sodium Carbonate 366
Sodium Chloride 367
Sodium Fluoride 368
Sodium Hydroxide 369
Sodium Hypochlorite 370
Sodium Hypophosphite 371
Sodium Metasilicate 372
Sodium Nitrate 373
Sodium Nitrite 374
Sodium Sulfate 375
Sodium Sulfite 376
Sodium Tetraborate 377
Sodium Thiosulfate 378
Solder Flux, Rosin 379
Stannic Chloride 380

Stannous Chloride **381**
Stannous Sulfate **382**
Stoddard Solvent **383**
Styrene, Monomer **384**
Succinic Acid **385**
Sulfamic Acid **386**
Sulfur Hexafluoride **387**
Sulfuric Acid **388**
Sulfurous Oxychloride 396

T

TCA 396, 402, 404
TCE 405
TDI 401
TEA 407
TEG 408
Tellurium **389**
TEOS 393
Tergitol **390**
Tetrachlorobutadiene **391**
Tetrachloroethylene **392**
Tetrachloromethane 167
Tetraethyl Ortho Silicate **393**
Tetraethyl Silicate 393
Tetrafluoromethane 168
Tetrafluorosilane 358
Tetrahydrofuran **394**
Tetramethyl Ammonium Hydroxide **395**
THF 394
Thionyl Chloride **396**
Thiourea **397**
Tin, Inorganic **398**
Tin Chloride 380
Tin Dichloride 381
Tin Sulfate 382
Titanium **399**
Toluene **400**
Toluene-2,4-Diisocyanate **401**
Toluol 400

Trichloroacetic Acid **402**
Trichlorobenzene **403**
1,1,1-Trichloroethane **404**
Trichloroethylene **405**
Trichloromethane 172
Trichlorosilane **406**
Triethanolamine **407**
Triethylene Glycol **408**
Triglycol 408
Trimethylaluminum **409**
Trimethylantimony **410**
Trimethylarsenic **411**
Trimethyl Borate **412**
Trimethyl Phosphite **413**
Trisodium Phosphate **414**
Trisodium Phosphate Dodecahydrate 414
TSP 414
Tungsten **415**
Tungsten Hexafluoride **416**

V

Vinyl Cyclohexene Dioxide **417**
Vinylhexane Dioxide 417

W

White Arsenic 135
Wood Alcohol 285

X

Xenon 258
Xylene **418**
Xylol 418

Z

Zinc Oxide **419**

NOTES:

NOTES:

NOTES:

NOTES:

NOTES:

NOTES:

NOTES:

NOTES:

NOTES: